Timing

TIMING

by

Sachin Sapatnekar
University of Minnesota, U.S.A.

KLUWER ACADEMIC PUBLISHERS
Boston / Dordrecht / New York / London

Distributors for North, Central and South America:
Kluwer Academic Publishers
101 Philip Drive
Assinippi Park
Norwell, Massachusetts 02061 USA
Telephone (781) 871-6600
Fax (781) 871-6528
E-Mail <kluwer@wkap.com>

Distributors for all other countries:
Kluwer Academic Publishers Group
Post Office Box 322
3300 AH Dordrecht, THE NETHERLANDS
Telephone 31 78 6576 000
Fax 31 78 6576 474
E-Mail <orderdept@wkap.nl>

Electronic Services <http://www.wkap.nl>

Library of Congress Cataloging-in-Publication

Title: Timing
Author (s): Sachin Sapatnekar
ISBN: 1-4020-7671-1
ISBN: 1-4020-8022-0 (eBook)

To Ofelia, from whom I've
learnt that timing isn't
everything

Contents

Acknowledgements

I would like to acknowledge the help and support of several people who have supported this effort. Specific and valuable feedback that has helped to orient this book at various levels has come from Haihua Su, David Blaauw, and Chandramouli Kashyap.

Several of my students have helped develop portions of this book during their studies, including Hongliang Chang, Rahul Deokar, Jiang Hu, Yanbin Jiang, Kishore Kasamsetty, Mahesh Ketkar, Daksh Lehther, Naresh Maheshwari, Harsha Sathyamurthy, Vijay Sundararajan, and Min Zhao. Collaborations and other interactions over the years with Chuck Alpert, Weitong Chuang, Madhav Desai, Jack Fishburn, Noel Menezes, Sani Nassif, Rajendran Panda, Darshan Patra, and Chandu Visweswariah have helped to add material to this book. Haihua Su and Shrirang Karandikar have greatly helped in proofreading parts of the book, as has my EE 5302 class in the Spring of 2004. My graduate students have been very forgiving of my tardiness in responding to them over the last year, and I am obliged to them for their patience.

I would like to convey my appreciation to Vasant Rao, Steve Kang, Ibrahim Hajj and Res Saleh at Illinois, Vijay Pitchumani at Syracuse and H. Narayanan at IIT Bombay, for introducing me to this area. This book is influenced by their classes, and by other conversations with them.

This book has been made possible by the research support of many agencies, primarily the National Science Foundation and the Semiconductor Research Corporation.

I owe a debt of gratitude to my parents, Suresh and Sudha Sapatnekar, and my sister, Suneeti. And, of course, to Ofelia.

Finally, I would like to thank all of my teachers and all of my students (who have been among my best teachers).

April 19, 2004   SSS

1 PREDUCTION/INTROFACE

It is customary to begin a book of this sort with Deep and Noble Thoughts. There are a number of excellent sources that convincingly make the point that timing is extremely important in current and future technologies, and as a Bear of Very Little Brain [Mil28], the author has little to add to this. Therefore, all that there is to say is presented in a hybrid between a preface and an introduction (which, in case you were wondering, is where the title of this chapter came from). Suffice it to say that if you've picked up this book, you're probably interested in timing in one way or another.

This book attempts to provide an overview of methods that are used for the analysis and optimization of timing in digital integrated circuits in contemporary technologies. It is rather well known that this topic is important to circuit designers and CAD engineers. The problem of timing analysis and optimization has come a long way over the years, and has become increasingly complicated with the passage of time due to two classes of effects that arise due to shrinking device geometries:

"Micro" effects correspond to new physical effects at the device or interconnect level caused by reductions in device dimensions.

"Macro" effects are related to problems of scale, as circuit sizes increase with the ability to pack larger circuits on a die.

The former necessitates novel approaches that handle the new effects, while the latter requires great efficiency in the algorithms to handle increasing problem sizes.

One of the holy grails of design is to achieve timing closure in a single pass of design; simply put, this implies that the design should meet its specifications without iteration. This is a rather difficult task, and a more realistic goal is to minimize the number of iterations by using realistic timing information at every step of design. Unfortunately, this is limited by the fact that early design stages, which show great flexibility towards design change, must operate under incomplete layout information, whereas later steps, where the more detailed layout provides accurate timing estimates, show a lower degree of flexibility.

Developing accurate methods for timing is, therefore, a vital part of ensuring fast timing closure. As a part of this goal, some kind of timing analysis must be carried out at each stage of design, right from a behavioral or register transfer level (RTL) description, down through logic synthesis, transistor-level optimization, and layout, working with the level of available information at each stage. This book primarily focuses on timing issues that are used in the design cycle in the post-synthesis phase, although some of the techniques may be used at earlier steps of the design cycle.

The organization of the book is as follows:

- In Chapter 2, we go through a quick tour of circuit simulation, which should be just enough to introduce the concepts that are required for fast timing analysis.

- Next, in Chapter 3, fast techniques for the analysis of linear systems using model order reduction and related techniques are studied.

- Since real circuits tend to have a mix of linear and nonlinear elements, we stretch this linear world to include nonlinear elements in Chapter 4, where we discuss methods for finding the delay of a single stage of digital logic.

- This is then expanded to the static timing analysis of full combinational circuits in Chapter 5, incorporating the effects of transistor nonlinearities with the linear behavior of interconnect parasitics.

- All of the analyses up to this point have been purely deterministic, but the increasing amount of uncertainty in upcoming designs has motivated the need for statistical timing analysis. This is a relatively recent field, and an overview of approaches for statistical static timing analysis that have been proposed to date is presented in Chapter 6.

- The next step is to overview techniques for the analysis of edge-triggered and level-clocked sequential circuits in Chapter 7.

- Chapter 8 then presents a few techniques for the transistor-level optimization of combinational circuits, specifically using the techniques of transistor sizing and dual V_t assignment for purposes of illustration.

- Chapter 9 considers the effects of intentional or deliberate skews in the clock network, surveying techniques for clock network construction, zero and nonzero skew optimization, and combinational optimization in conjunction with the use of deliberate skews.

- The technique of retiming, a purely sequential optimization method, is introduced in Chapter 10, and algorithmic techniques for retiming are discussed.

- Finally, the book closes out with some concluding remarks in Chapter 11.

It is unlikely that this book is free of errors, and a list of errata will be maintained at

`http://www.ece.umn.edu/users/sachin/timingbook`

2 A QUICK OVERVIEW OF CIRCUIT SIMULATION

2.1 INTRODUCTION

Circuit simulation provides an methodical procedure for writing the equations that describe a circuit, and utilizes numerical techniques to solve them to an accuracy within the limits of the numerical methods and machine precision. The result of this computation can be used to determine the currents and voltages throughout the circuit over a specified time period of interest. It is generally considered to be the most precise practical way of solving general circuits. In this chapter, we will overview the techniques used by circuit simulators such as SPICE [Nag75]. This discussion aims to be sufficient to acquaint the reader with the background required for understanding concepts in timing analysis, and is not intended to be completely thorough; for such coverage, the reader is referred to [CL75, VS94, PRV95].

At the most elementary level, a circuit may consist of fundamental elements such as resistors, capacitors, inductors, transistors, diodes and controlled sources. A circuit is an interconnection of these fundamental elements, and the currents and voltages in the circuit are governed by three basic sets of equations:

- Kirchoff's Current Law (KCL), which states that the algebraic sum of currents leaving any node is zero,

- Kirchoff's Voltage Law (KVL), which affirms that the current around any cycle in a circuit must be zero, and

- the device equations that relate the current and voltage parameters for individual elements.

The KCL and KVL relations are *linear* equations that are purely topological in that they depend only on the connectivity of the network, and not on the type of elements in the circuit. The element characteristics are wholly modeled by the device equations, and these may, in general be *linear* or *nonlinear* or *differential* equations. Circuit simulation involves the solution of this system of equations to find the currents and voltages everywhere in the circuit. For practical purposes, the results of circuit simulation are often considered to be the exact solution to any circuit, although it is useful to temper this statement with the observation that any such solution is only as exact as the models that are used for the devices.

For a circuit with n nodes (plus a ground node) and b branches, there are n independent KCL equations, b independent KVL equations, and b device equations. The circuit variables consist of n node voltages, b branch currents and b branch voltages, which yields a total of $2b + n$ equations in an equal number of variables.

There are several ways of combining these equations into a more compact set, but we will focus our attention on the *modified nodal analysis* (MNA) method since it is arguably the most widely-used formulation.

2.2   FORMULATION OF CIRCUIT EQUATIONS

We first introduce a formal manner for writing the equations for a circuit. To write the topological KCL and KVL equations, we will consider a circuit to be represented by a directed graph $G = (V, E)$, where the vertex set V has a one-to-one correspondence with the set of nodes in the circuit, and the elements in the edge set E correspond to the branches. The directions on the edges may be chosen arbitrarily[1]. Given any such graph, we can define the notion of an *incidence matrix*, A_{inc}, on the graph G. This is an $(n + 1) \times b$ matrix, with the rows corresponding to the n nonground vertices and one vertex corresponding to the ground node. Each of the b columns corresponds to a directed edge and has two nonzero entries: a "+1" for the source of a directed edge, and a "-1" for the destination. An example incidence matrix for a sample circuit graph is shown in Figure 2.1. The reference directions chosen for the branches are shown by arrows in the circuit graph.

It may be observed that by definition, the sum of all entries in each column of A_{inc} is 0; as a consequence, A_{inc} is not of full rank[2]. It can be shown that for a connected graph G[3], the rank of A_{inc} is n, and that any $n \times n$ submatrix of A_{inc} is nonsingular. In other words, eliminating one row of A_{inc} (typically the row corresponding to the ground node) would convert it to a matrix of full rank that we will denote as A.

Let $\mathbf{I_b}$ denote the vector of branch currents. With the aid of the incidence matrix, the n KCL equations may be written as follows:

$$A \cdot \mathbf{I_b} = \mathbf{0} \qquad (2.1)$$

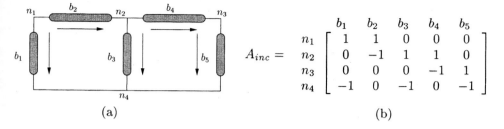

Figure 2.1. (a) A circuit with five branches and four nodes and (b) its incidence matrix. The reference directions for the branches are arbitrarily chosen, and are shown by the arrows.

The explanation for this is simple. Since each row of A has a "+1" entry corresponding to a branch that has its source at that node, and a "-1" for a branch terminating at that node, each row of the matrix product $A \cdot \mathbf{I_b}$ is the algebraic sum of the currents leaving the node, which equals zero.

If $\mathbf{V_b}$ is the vector of branch voltages and $\mathbf{V_n}$ the vector of node voltages, then it is easy to see that the two are related by the b equations

$$\mathbf{V_b} = A^T \cdot \mathbf{V_n} \tag{2.2}$$

From the definition of A, we see that each row in the above equation corresponds to a statement that the voltage across a branch is given by the difference between the voltages at either end. Although it may not be immediately obvious, this is indeed the set of KVL equations. By way of verification, the reader may add the voltages around any cycle and will find that the variables cancel telescopically to sum up to zero. For instance, the application of the above equation to the sum of the branch voltages in the cycle (b_1, b_2, b_3) in Figure 2.1 yields

$$
\begin{aligned}
\text{Voltage across the cycle } (b_1, b_2, b_3) \quad &= \quad -V_{b_1} + V_{b_2} + V_{b_3} \\
&= \quad -V_{n_1} + V_{n_4} + V_{n_1} - V_{n_2} + V_{n_2} - V_{n_4} \\
&= \quad 0
\end{aligned}
$$

Finally, the device equations describe the relation between the $\mathbf{V_b}$ and $\mathbf{I_b}$ vectors. These equations could be linear or nonlinear or differential equations, and for the time being, we will focus our attention on devices whose characteristics are defined by linear algebraic equations. For convenience, we will divide these into two categories:

Type 1 devices can be defined by a device equation of the type

$$\mathbf{I_b} = Y\mathbf{V_b} + \mathbf{s} \tag{2.3}$$

If b_1 is the total number of such devices in the circuit, then Y is a constant $b_1 \times b_1$ matrix and $\mathbf{s}$ is a constant $b_1 \times 1$ vector. Examples of such devices are conductances, constant current sources, and voltage-controlled current sources.

Type 2 devices are those that are not described by Type 1 equations, and are instead described by equations of the type

$$Z\mathbf{I_b} + G\mathbf{V_b} = \mathbf{t} \tag{2.4}$$

If b_2 is the total number of Type 2 devices in the circuit, then Z and G are constant $b_2 \times b_2$ matrices, and $\mathbf{t}$ is a constant $b_2 \times 1$ vector. Constant voltage sources and current-controlled current/voltage sources cannot be described by Type 1 equations, and therefore they are examples of this class of devices.

Let us now consider a circuit that contains only Type 1 and Type 2 devices. Clearly, for such a circuit, the sum of b_1 and b_2 is b. The $2b+n$ relations describing this system correspond to Equations (2.1), (2.2), (2.3) and (2.4). However, simple algebraic substitutions may be used to reduce these to a smaller set of equations.

For convenience, we will order the branch voltage and branch current vectors so that the Type 1 devices are listed before the Type 2 devices. The incidence matrix is also correspondingly rearranged, with the submatrix corresponding to Type 1 [Type 2] devices being referred to as A_1 [A_2]. We will denote the vector of currents and voltages corresponding to Type 1 [Type 2] elements by $\mathbf{I_{b_1}}$ and $\mathbf{V_{b_1}}$ [$\mathbf{I_{b_2}}$ and $\mathbf{V_{b_2}}$], respectively. The list of all circuit equations is then:

$$KCL: \quad \left[A_1 \vdots A_2 \right] \begin{bmatrix} \mathbf{I_{b_1}} \\ \cdots \\ \mathbf{I_{b_2}} \end{bmatrix} = \mathbf{0}$$

$$\Rightarrow \quad A_1\mathbf{I_{b_1}} + A_2\mathbf{I_{b_2}} = \mathbf{0} \tag{2.5}$$

$$KVL: \quad \begin{bmatrix} \mathbf{V_{b_1}} \\ \cdots \\ \mathbf{V_{b_2}} \end{bmatrix} = \left[A_1 \vdots A_2 \right]^T \mathbf{V_n}$$

$$\Rightarrow \quad \mathbf{V_{b_1}} = A_1^T\mathbf{V_n} \tag{2.6}$$

$$\text{and} \quad \mathbf{V_{b_2}} = A_2^T\mathbf{V_n} \tag{2.7}$$

$$\textit{Device Equations (Type 1):} \quad \mathbf{I_{b_1}} = Y\mathbf{V_{b_1}} + \mathbf{s} \tag{2.8}$$

$$\textit{Device Equations (Type 2):} \quad Z\mathbf{I_{b_2}} + G\mathbf{V_{b_2}} = \mathbf{t} \tag{2.9}$$

Combining (2.5), (2.6), and (2.8), we obtain

$$A_1 Y A_1^T \mathbf{V_n} + A_1\mathbf{s} + A_2\mathbf{I_{b_2}} = \mathbf{0} \tag{2.10}$$

The two remaining equations may also be merged: by inserting (2.7) into (2.9), we get

$$Z\mathbf{I_{b_2}} + G A_2^T \mathbf{V_n} = \mathbf{t} \tag{2.11}$$

The equations (2.10) and (2.11) may be written together in terms of the *modified nodal formulation* (MNA) as:

$$
\begin{bmatrix} A_1 Y A_1^T & \vdots & A_2 \\ \cdots\cdots\cdots\cdots & & \\ G A_2^T & \vdots & Z \end{bmatrix}
\begin{bmatrix} \mathbf{V_n} \\ \cdots \\ \mathbf{I_{b_2}} \end{bmatrix}
=
\begin{bmatrix} -A_1 \mathbf{s} \\ \cdots \\ \mathbf{t} \end{bmatrix}
\tag{2.12}
$$

We observe that the variables in the MNA equation correspond to the voltages at each of the nodes and the currents through the Type 2 elements, for a total of $n + b_2$ equations in as many variables. When all elements are of Type 1, this set of equations is referred to as the *nodal formulation*.

2.2.1   Stamps for commonly-encountered elements

It may appear that finding the above expression would require a good deal of computation, involving the multiplication of large matrices. Fortunately, there are simple techniques for finding the entries of the MNA matrix by developing "stamps" for individual elements, and formulating the left hand side matrix and right hand side vector by inspection of a circuit. To illustrate this, we will develop stamps for a few typical elements:

Conductances.   For a conductance g connected on a branch b_g between nodes a and b, and carrying a current I_g and a voltage V_g, each with a reference direction from a to b, the contribution to the KCL equation can be shown in terms of its entries in the incidence matrix as follows:

$$
\begin{array}{c}
 \\
a \\
 \\
b
\end{array}
\begin{bmatrix}
b_g \\
+1 \\
-1
\end{bmatrix}
\begin{bmatrix}
I_g
\end{bmatrix}
= 0
\tag{2.13}
$$

Substituting the device equation for the conductance, $I_g = g V_g$, we can rewrite the above equation as

$$
\begin{array}{c}
 \\
a \\
 \\
b
\end{array}
\begin{bmatrix}
b_g \\
+g \\
-g
\end{bmatrix}
\begin{bmatrix}
V_g
\end{bmatrix}
= 0
\tag{2.14}
$$

Finally, the only remaining relation to be substituted is the KVL equation that states that $V_g = V_a - V_b$, where V_a and V_b are, respectively, the voltages at the nodes a and b. We obtain

$$
\begin{array}{c} \\ a \\ \\ b \end{array}
\begin{array}{cc} a & b \end{array}
\left[\begin{array}{cc} +g & -g \\ -g & +g \end{array} \right]
\left[\begin{array}{c} V_a \\ V_b \end{array} \right] = \mathbf{0}
\qquad (2.15)
$$

This implies that for a conductance connected between nodes a and b, there are no contributions to the RHS matrix, and the LHS matrix has only four entries: $+g$ at the (a, a) and (b, b) positions, and $-g$ at the (a, b) and (b, a) positions. It may be observed that since a conductance is a Type 1 element, the only contributions are to the $A_1 Y A_1^T$ region of the MNA matrix.

Constant current sources. Now consider another Type 1 element: a constant current source J on a branch b_J, directed from node a to node b. Its contribution to the KCL equation is

$$
\begin{array}{c} a \\ \\ b \end{array}
\begin{array}{c} b_J \end{array}
\left[\begin{array}{c} +1 \\ -1 \end{array} \right]
\left[\begin{array}{c} I_g \end{array} \right] =
\left[\begin{array}{c} \\ \end{array} \right]
\qquad (2.16)
$$

Substituting the device equation for the conductance, $I_g = J$, we can find its stamp as

$$
\left[\begin{array}{c} \\ \end{array} \right]
\left[\begin{array}{c} \\ \end{array} \right] =
\begin{array}{c} a \\ \\ b \end{array}
\left[\begin{array}{c} -J \\ +J \end{array} \right]
\qquad (2.17)
$$

The stamp for a constant current source is, therefore, an entry of $-J$ in the RHS matrix in the ath position and $+J$ in the bth position.

Constant voltage sources. A constant voltage source of value V connected on a branch b_V between nodes a and b is an example of a Type 2 element. Therefore, its contributions to the KCL equation are similar to the elements above; the difference here is that the branch current variable remains present in the final set of MNA equations.

The Type 2 device equation, combined with the KVL equation, states that $V_a - V_b = V$. This is written in the lower part of the MNA matrix, after all the KCL equations, as follows:

$$
\begin{array}{c}
a \\
b \\
b_V
\end{array}
\begin{bmatrix}
& & \vdots & +1 \\
& & \vdots & -1 \\
\cdots\cdots\cdots\cdots\cdots \\
+1 & -1 & \vdots &
\end{bmatrix}
\overset{b_V}{}
\begin{bmatrix}
V_a \\
V_b \\
\cdots \\
I_b
\end{bmatrix}
=
\begin{bmatrix}
\\
\\
V
\end{bmatrix}
\tag{2.18}
$$

Other elements.   The stamps for several other elements are as shown below; their derivation is left as an exercise to the reader.

Capacitors in the Laplace domain   A capacitor of value C is represented as an admittance of sC. Its stamp can therefore be derived to be identical to the stamp for the conductance shown in Equation (2.15), with g replaced by sC, to give:

$$
\begin{array}{c}
a \\
b
\end{array}
\overset{\begin{array}{cc} a & \quad b \end{array}}{
\begin{bmatrix}
+sC & -sC \\
-sC & +sC
\end{bmatrix}}
\begin{bmatrix}
V_a \\
V_b
\end{bmatrix}
=
\begin{bmatrix}
\\
\end{bmatrix}
\tag{2.19}
$$

Inductors in the Laplace domain   A similar argument may be used to state that the stamp for a self-inductance of value L is the same as that for the conductance, but with g replaced by $1/sL$. However, for reasons of convenience that we will further elaborate on in Section 3.5.2, inductors are often represented as Type 2 elements, so that their stamps can be derived, in a manner similar to the derivation of Equation (2.18), as

$$
\begin{array}{c}
a \\
b \\
b_L
\end{array}
\overset{\begin{array}{ccc} a & \; b & \; b_L \end{array}}{
\begin{bmatrix}
& & \vdots & +1 \\
& & \vdots & -1 \\
\cdots\cdots\cdots\cdots\cdots\cdots \\
+1 & -1 & \vdots & -sL
\end{bmatrix}}
\begin{bmatrix}
V_a \\
V_b \\
\cdots \\
I_b
\end{bmatrix}
=
\begin{bmatrix}
\\
\\
\end{bmatrix}
\tag{2.20}
$$

For a system of mutual inductances, in the Type 2 device equation, $G = I$, where I is the identity matrix, and $Z = sM$. Therefore, the corresponding stamp is given by:

$$
\begin{bmatrix}
& \vdots & A_2 \\
\cdots\cdots\cdots\cdots \\
A_2^T & \vdots & -sM
\end{bmatrix}
\begin{bmatrix}
\mathbf{V_n} \\
\cdots \\
\mathbf{I_M}
\end{bmatrix}
=
\begin{bmatrix}
\\
\end{bmatrix}
\tag{2.21}
$$

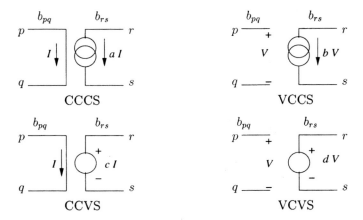

Figure 2.2.   Various types of controlled sources.

Controlled sources For the controlled sources shown in Figure 2.2, the MNA stamps are as follows:

Current-controlled current source (CCCS):

$$
\begin{bmatrix}
 & p & q & & b_{pq} & \\
p & & & : & +1 & \\
q & & & : & -1 & \\
 & & & : & & \\
r & & & : & +a & \\
s & & & : & -a & \\
 & & & : & & \\
\hline
b_{pq} & +1 & -1 & : & &
\end{bmatrix}
\begin{bmatrix}
V_p \\
V_q \\
\\
V_r \\
V_s \\
\cdots \\
I_{pq}
\end{bmatrix}
=
\begin{bmatrix}
\\
\\
\\
\\
\\
\\
\\
\end{bmatrix}
\tag{2.22}
$$

Voltage controlled current source (VCCS)

$$
\begin{bmatrix}
 & p & q & r & s \\
p & & & & \\
q & & & & \\
r & +b & -b & & \\
s & -b & +b & &
\end{bmatrix}
\begin{bmatrix}
V_p \\
V_q \\
V_r \\
V_s
\end{bmatrix}
=
\begin{bmatrix}
\\
\\
\\
\end{bmatrix}
\tag{2.23}
$$

Note that this element does not have any branches that are classified as Type 2 elements. While this stamp looks superficially similar to that for the conductance, it differs in that it is asymmetric: it contains no entries in the rows of p and q, or in the columns r and s, and will lead to an asymmetric nodal or MNA matrix.

Current controlled voltage source (CCVS)

$$
\begin{array}{c}
\\
p \\
q \\
\\
r \\
s \\
\\
b_{pq} \\
b_{rs}
\end{array}
\begin{bmatrix}
 & & & & & b_{pq} & b_{rs} \\
 & & & & \vdots & +1 & \\
 & & & & \vdots & -1 & \\
 & & & & \vdots & & \\
 & & & & \vdots & & +1 \\
 & & & & \vdots & & -1 \\
\cdots & \cdots & \cdots & \cdots & \cdots & \cdots & \cdots \\
+1 & -1 & & & \vdots & & \\
 & & +1 & -1 & \vdots & -c &
\end{bmatrix}
\begin{bmatrix}
V_p \\
V_q \\
\\
V_r \\
V_s \\
\cdots \\
I_{pq} \\
I_{rs}
\end{bmatrix}
=
\begin{bmatrix}
\; \\
\; \\
\; \\
\; \\
\; \\
\; \\
\;
\end{bmatrix}
\tag{2.24}
$$

Voltage controlled voltage source (VCVS)

$$
\begin{array}{c}
\\
\\
r \\
s \\
\\
\\
b_{pq}
\end{array}
\begin{bmatrix}
 p & q & & r & s & b_{pq} \\
 & & & & & \vdots \\
 & & & & & \vdots \\
 & & & & & \vdots & +1 \\
 & & & & & \vdots & -1 \\
 & & & & & \vdots \\
 & & & & & \vdots \\
\cdots & \cdots & \cdots & \cdots & \cdots & \cdots \\
-d & +d & & +1 & -1 & \vdots
\end{bmatrix}
\begin{bmatrix}
V_p \\
V_q \\
\\
V_r \\
V_s \\
\\
\\
I_{rs}
\end{bmatrix}
=
\begin{bmatrix}
\; \\
\; \\
\; \\
\; \\
\; \\
\;
\end{bmatrix}
\tag{2.25}
$$

2.3   EXAMPLES OF EQUATION FORMULATION BY INSPECTION

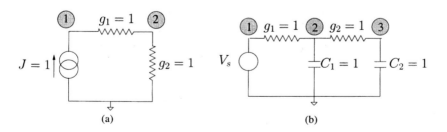

Figure 2.3.   Circuit examples to illustrate the MNA formulation.

Example 1: Consider the circuit shown in Figure 2.3(a) with two unit resistors, driven by a unit current source. The circuit has two non-ground nodes and no

Type 2 elements, and therefore its equations can be written using the nodal formulation. The circuit is described by a system of two equations in the two nodal variables, V_1 and V_2, the voltages at the two non-ground nodes. Adding up the contributions of the conductance stamp from Equation (2.15) and the current source stamp from Equation (2.17), the nodal equations are found to be:

$$\begin{bmatrix} 1 & \vdots & -1 \\ \cdots\cdots\cdots\cdots \\ -1 & \vdots & 1+1 \end{bmatrix} \begin{bmatrix} V_1 \\ \cdots \\ V_2 \end{bmatrix} = \begin{bmatrix} 1 \\ \cdots \\ 0 \end{bmatrix}$$

Note that the conductance g_1 contributes to all four elements of the LHS matrix, g_2 to only the (2,2) element (since its other end is connected to ground), and, for the same reason, J to only one entry on the RHS.

It is easily seen that the solution to this system is $V_1 = 2, V_2 = 1$, which is the expected solution for such a resistive divider. □

Example 2: The circuit in Figure 2.3(b) corresponds to Figure 2.1(a), where the elements in branches b_2 and b_4 are unit resistances, those in b_3 and b_5 are unit capacitances, and the element b_1 is a voltage source excitation, $V(s)$, with a directionality from n_1 to n_4; the latter is taken to be the ground node. This system contains one Type 2 element, the voltage source, and three nodal voltage variables. In the frequency (Laplace) domain, the MNA equations can therefore be written as

$$\begin{bmatrix} 1 & -1 & 0 & 1 \\ -1 & 2+s & -1 & 0 \\ 0 & -1 & 1+s & 0 \\ 1 & 0 & 0 & 0 \end{bmatrix} \begin{bmatrix} V_1 \\ V_2 \\ V_3 \\ I_v \end{bmatrix} = \begin{bmatrix} 0 \\ 0 \\ 0 \\ V_s \end{bmatrix}$$

□

2.4   SOLUTION OF NONLINEAR EQUATIONS

In Section 2.2, an equation formulation technique for circuits described by linear algebraic equations was described. In this section, we will build upon this foundation to consider how devices that are described by nonlinear algebraic equations may be incorporated into such a formulation.

To motivate this idea, let us consider the example of a diode connected between nodes a and b. This element is represented by the equation

$$i = I_o(e^{\frac{v}{V_t}} - 1) \tag{2.26}$$

where i and v are, respectively, the current through and the voltage across the diode, and V_t is the thermal voltage, which is the product of a physical constant and the temperature. These device characteristics are pictorially illustrated by the curve shown in Figure 2.4. The operating point (i, v) of the diode depends on the device equations for the other elements of the circuit and on

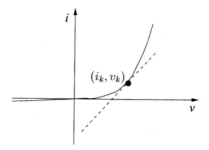

Figure 2.4.   The device equation of a diode and a local linear approximation about (i_k, v_k).

the circuit topology. Let us assume that we iteratively proceed to find this value, beginning with an initial guess. In the kth iteration, let us assume that this value is (i_k, v_k). We may then approximate the diode equation by a tangent to the curve at this point, as shown in the figure. This tangent provides a local linear approximation[4] to the nonlinear curve. Specifically, this may be obtained through a first order Taylor series expansion as

$$i \;=\; i_k + \left.\frac{\partial i}{\partial v}\right|_{(i_k, v_k)} (v - v_k)$$

$$\Rightarrow i \;=\; I_o(e^{\frac{v_k}{V_t}} - 1) + \frac{I_o}{V_t}e^{\frac{v_k}{V_t}}(v - v_k)$$

$$\Rightarrow i \;=\; a_k v + b_k \tag{2.27}$$

where

$$a_k \;=\; \frac{I_o}{V_t}e^{\frac{v_k}{V_t}}$$

$$b_k \;=\; I_o(e^{\frac{v_k}{V_t}} - 1) - \frac{I_o}{V_t}e^{\frac{v_k}{V_t}}v_k$$

It is easy to see that this local linear approximation is a Type 1 equation; in particular, it is equivalent to a conductance of a_k in parallel with a current source of value b_k. Therefore, it may be represented in the circuit by adding the stamps for these elements, so that the composite stamp for a diode in the kth iteration is

$$\begin{bmatrix} a_k & -a_k \\ \\ -a_k & a_k \end{bmatrix} \begin{bmatrix} V_a \\ \\ V_b \end{bmatrix} = \begin{bmatrix} -b_k \\ \\ b_k \end{bmatrix} \tag{2.28}$$

In the equation above, V_a and V_b are MNA variables that correspond to the voltages at the two nodes a and b between which the diode is connected.

Having observed this principle, let us now consider a device equation of the type

$$f(x_1, x_2, \cdots, x_n) = 0 \qquad (2.29)$$

Such an equation may be approximated using a truncated first order Taylor series about a point $\mathbf{x}_k = (x_1^k, x_2^k, \cdots, x_n^k)$ (we will see later that the subscript k corresponds to an iteration number). This results in the following equation:

$$f(x_1^k, x_2^k, \cdots, x_n^k) + \sum_{i=1}^{n} \left[\frac{\partial f(x_1, x_2, \cdots, x_n)}{\partial x_i} \bigg|_{\mathbf{x}_k} (x_i - x_i^k) \right] = 0 \qquad (2.30)$$

Note that the partial derivative is evaluated at $\mathbf{x}_k$, so that this results in a linear approximation. An alternative interpretation of this is that it approximates the nonlinear function by its tangent plane at the point $\mathbf{x}_k$ to arrive at a local linear approximation. This procedure is identical to the Newton-Raphson method that is used to find the roots of a nonlinear system of equations, and is often referred to by that name (or simply as "Newton's method").

Returning to the original context, consider a circuit consisting only of devices whose characteristic equation may be represented by linear or nonlinear algebraic equations. Recall that both KCL and KVL are linear. If the nonlinear equations were to be in the form of Equation (2.29), then the locally linear approximation provided by Equation (2.30) could be used to obtain a system of purely linear algebraic equations. Further, any such linearized device equation could be taken to be a Type 1 or Type 2 device, which implies that the circuit could be represented using the MNA formulation.

In general, such a locally linear approximation is accurate only within a small region around the expansion point $\mathbf{x}_k$; if the actual operating point of the device is within this region, then the solution to the linearized MNA formulation is a valid solution to the original nonlinear system. If it is not, then the MNA solution will lead to a new solution point, $\mathbf{x}_{k+1}$. A new guess for the operating point may be obtained using this value, and the circuit may be linearized about the new operating point. This process is repeated until convergence.

For a general system of equations, there is no guarantee of convergence if we start from an arbitrary initial guess. However, in practice, it is often possible to obtain reasonable initial guesses, particularly for digital circuits, which allows the procedure to converge relatively soon.

Example: A Level 1 SPICE model for an nmos transistor is given by the equations

$$I_{ds} = \begin{cases} 0 & V_{gs} \leq V_t & \text{(cutoff)} \\ K\left((V_{gs} - V_t)V_{ds} - \frac{V_{ds}^2}{2})\right) & 0 \leq V_{ds} \leq V_{gs} - V_t & \text{(linear)} \\ K\frac{(V_{gs} - V_t)^2}{2}(1 + \lambda V_{ds}) & 0 \leq V_{gs} - V_t \leq V_{ds} & \text{(saturation)} \end{cases} \qquad (2.31)$$

where I_{ds} is the current from drain to source, V_{gs} and V_{ds} are the gate-to-source and drain-to-source voltages, respectively, V_t is the threshold voltage, and K is a constant that depends on the dimensions of the transistor and the physical properties of silicon. For simplicity, we will ignore the body effect and consider the device to be a three-terminal structure.

During the kth Newton iteration, the linearized equations for the nmos device about the point $(V_{gs}^k, V_{ds}^k, I_{ds}^k)$ are given by:

$$I_{ds} = \begin{cases} 0 & \text{(cutoff)} \\ I_{ds}^k + \alpha_{k,lin}(V_{gs} - V_{gs}^k) + \beta_{k,lin}(V_{ds} - V_{ds}^k) & \text{(linear)} \\ I_{ds}^k + \alpha_{k,sat}(V_{gs} - V_{gs}^k) + \beta_{k,sat}(V_{ds} - V_{ds}^k) & \text{(saturation)} \end{cases} \qquad (2.32)$$

where

$$\alpha_{k,lin} = \left.\frac{\partial I_{ds,linear}}{\partial V_{gs}}\right|_k = K V_{ds}^k$$

$$\beta_{k,lin} = \left.\frac{\partial I_{ds,linear}}{\partial V_{ds}}\right|_k = K(V_{gs}^k - V_t - V_{ds}^k)$$

$$\alpha_{k,sat} = \left.\frac{\partial I_{ds,saturation}}{\partial V_{gs}}\right|_k = K(V_{gs}^k - V_t)(1 + \lambda V_{ds}^k)$$

$$\beta_{k,sat} = \left.\frac{\partial I_{ds,saturation}}{\partial V_{ds}}\right|_k = K\lambda\frac{(V_{gs}^k - V_t)^2}{2}$$

The cutoff, linear and saturation regions are as defined in Equation (2.31). □

Let us now present a general framework for handling a circuit described by linear and nonlinear algebraic equations. Consider a circuit that contains b devices, of which the characteristic equations of p devices are represented by nonlinear algebraic equations; the remaining $b - p$ characteristic equations are linear algebraic equations. We will write the p nonlinear equations as

$$\mathbf{f}(\mathbf{x}_k) = \begin{bmatrix} f_1(x_1, \cdots, x_k) \\ f_2(x_1, \cdots, x_k) \\ \vdots \\ f_p(x_1, \cdots, x_k) \end{bmatrix} = \mathbf{0} \qquad (2.33)$$

We define the Jacobian for these equations as

$$J(\mathbf{x}_k) = \begin{bmatrix} \frac{\partial f_1}{\partial x_1} & \frac{\partial f_1}{\partial x_2} & \cdots & \frac{\partial f_1}{\partial x_n} \\ \frac{\partial f_2}{\partial x_1} & \frac{\partial f_2}{\partial x_2} & \cdots & \frac{\partial f_2}{\partial x_n} \\ \cdots & \cdots & \ddots & \cdots \\ \frac{\partial f_p}{\partial x_1} & \frac{\partial f_p}{\partial x_2} & \cdots & \frac{\partial f_p}{\partial x_n} \end{bmatrix} \qquad (2.34)$$

The approximating systems of equations may then be listed as

$$J(\mathbf{x}_k)\mathbf{x} = -\mathbf{f}(\mathbf{x}_k) + J(\mathbf{x}_k)\mathbf{x}_k \qquad (2.35)$$

which become Type 1 or Type 2 devices.

The computational overhead here consists of evaluating the Jacobian matrix in each iteration k, and this could be considerable. Several methods have been proposed to speed up this process. For instance, the parallel chord method assumes small perturbations from one iteration to the next, and updates the Jacobian only after several iterations.

The nonlinear iterations end when a convergence criterion is satisfied. One such convergence criterion is

$$||\mathbf{f}(\mathbf{x}_k)|| \le \epsilon \text{ and } ||\mathbf{x}_k - \mathbf{x}_{k-1}|| \le \delta \tag{2.36}$$

for appropriately chosen values of the tolerances ϵ and δ.

2.5 SOLUTION OF DIFFERENTIAL EQUATIONS

Now that our framework has been expanded to include any linear or nonlinear algebraic equation, let us consider devices that are represented by ordinary differential equations.

As a motivating example, let us consider an arbitrary device whose characteristics are given by the equation

$$\frac{dx}{dt} = f(x, t) \tag{2.37}$$

where x is the variable of interest and f is a general nonlinear equation. The initial condition for this differential equation is given as $x(t_0) = x_0$, where x_0 is a constant. Let us assume that we have calculated the value of x up to some arbitrary time point t_{n-1}, and let us consider how the next value at time point $t_n = t_{n-1} + h$ is computed, where h is referred to as the *time step*. It is enough to explain how we can compute x_{t_n} from this information; if we know this, we can start with $t_{n-1} = t_0$ and proceed one time step at a time, even possibly varying the time step from one time point to the next.

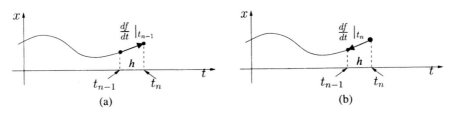

(a)   (b)

Figure 2.5. Numerical solution of a differential equation using (a) the Forward Euler method and (b) the Backward Euler method.

An intuitive way of performing this task may be understood visually by examining Figure 2.5(a). If we know the value of x and the time derivative of $\frac{dx}{dt}$ at time t_{n-1}, we can simply calculate the value of x at time t_n as

$$\begin{aligned} x_{t_n} &= x_{t_{n-1}} + h \left. \frac{dx}{dt} \right|_{t_{n-1}} \\ &= x_{t_{n-1}} + hf(x_{t_{n-1}}) \end{aligned} \tag{2.38}$$

This is referred to as the *Forward Euler* numerical integration formula, and it is merely a natural consequence of the finite difference approximation,

$$\frac{dx}{dt} = \frac{x_{t_n} - x_{t_{n-1}}}{h} \tag{2.39}$$

where the left hand side represents the derivative evaluated at time t_{n-1}.

Another alternative is to instead let the left hand side of Equation (2.39) correspond to the derivative at time t_n: this leads to the numerical integration formula,

$$
\begin{aligned}
x_{t_n} &= x_{t_{n-1}} + h \left. \frac{dx}{dt} \right|_{t_n} \\
&= x_{t_{n-1}} + h f(x_{t_n}) \tag{2.40}
\end{aligned}
$$

This technique, called the *Backward Euler* method may be interpreted as shown in Figure 2.5(b), where we take a step back from the time point t_n and use the derivative at that point to estimate the value there. Of course, the value of the derivative and the value x_{t_n} are both unknown, so that we obtain an implicit equation in x_{t_n} here, instead of the explicit equation provided by the Forward Euler method, where all terms on the right hand side are known. It can be seen then that both Equations (2.38) and (2.40) effectively convert the differential equation to a nonlinear equation at each time point.

The intuition behind yet another formula for numerical integration can likewise be arrived at. The Trapezoidal rule follows similar principles as the Forward Euler and Backward Euler methods, but averages the derivative at the points t_{n-1} and t_n, so that it may be stated as

$$
\begin{aligned}
x_{t_n} &= x_{t_{n-1}} + \frac{h}{2} \left(\left. \frac{dx}{dt} \right|_{t_{n-1}} + \left. \frac{dx}{dt} \right|_{t_n} \right) \\
&= x_{t_{n-1}} + \frac{h}{2} \left[f(x_{t_{n-1}}) + f(x_{t_n}) \right] \tag{2.41}
\end{aligned}
$$

As a motivating example to see how these techniques could be applied to circuit analysis, let us consider a linear capacitor of value C that is represented by the device equation

$$i = C \frac{dv}{dt} \tag{2.42}$$

where i and v are, respectively, the current through and voltage across the capacitor, and t is the time variable[5]. Let us assume that we know the value of this voltage and current at some time point t_{n-1}, and that we would like to calculate its value at the time point $t_n = t_{n-1} + h$ for a sufficiently small time step, h.

Using the Forward Euler method, we have

$$v_{t_n} = v_{t_{n-1}} + h \left. \frac{dv}{dt} \right|_{t_{n-1}} \tag{2.43}$$

Combining this with Equation (2.42) yields

$$v_{t_n} = v_{t_{n-1}} + \frac{h}{C}\dot{i}_{t_{n-1}} \tag{2.44}$$

Since the values at time t_n are unknown, we drop all subscripts n to obtain the approximated characteristic equation at t_n as

$$v = v_{t_{n-1}} + \frac{h}{C}\dot{i}_{t_{n-1}} \tag{2.45}$$

This corresponds to a constant voltage source, and therefore, the circuit may be analyzed by replacing the capacitor by a constant voltage source.

Alternatively, one could use the Backward Euler formula,

$$v_{t_n} = v_{t_{n-1}} + h \left.\frac{dv}{dt}\right|_{t_n} \tag{2.46}$$

Substituting the characteristic equation for the linear capacitor and dropping the t_n subscripts, we obtain the following approximation:

$$i = \frac{C}{h}v - \frac{C}{h}v_{t_{n-1}} \tag{2.47}$$

This corresponds to a conductance of value $\frac{C}{h}$ in parallel with a current source of value $-\frac{C}{h}v_{t_{n-1}}$. The stamps for these elements may be used to solve the circuit at time t_n.

If the Trapezoidal rule is used instead, it is easy to verify that for the linear capacitor, this results in the equation

$$i = \frac{2C}{h}v - \frac{2C}{h}v_{t_{n-1}} - i_{t_{n-1}}, \tag{2.48}$$

which again is a conductance in parallel with a current source.

All three methods use the information from one previous time step, but from Equations (2.47) and (2.48), we can see that the Backward Euler only requires the voltage across the capacitor at the previous time step, while the Trapezoidal Rule requires both the voltage and current from the previous time point.

Stability

While the preceding discussion makes these formulæ all look superficially similar, they have vastly different characteristics in terms of *numerical stability*. We explore the concept of numerical stability with the aid of a test equation,

$$\frac{dx}{dt} = \lambda x, x(t = 0) = x_0 \tag{2.49}$$

where λ is an arbitrary complex number. As $t \to \infty$, the solution $x(t)$ tends to zero when $Re(\lambda) < 0$, and it tends to ∞ when $Re(\lambda) > 0$. At the very minimum, the numerical solution should obey these limiting conditions, and if it does, we refer to the numerical integration formula as being stable.

Now consider the behavior of each of the above formulæ on this test equation, At the outset, we point out that since the step size $h > 0$, the sign of $Re(\lambda)$ is the same as that of $Re(h\lambda)$. For the Forward Euler method under a constant step size h, $\frac{dx}{dt}\big|_{x_i} = \lambda x_i \forall i$, so that

$$x_1 = x_0 + h \left.\frac{dx}{dt}\right|_{x_0} = (1 + h\lambda)x_0$$

$$x_2 = x_1 + h \left.\frac{dx}{dt}\right|_{x_1} = (1 + h\lambda)^2 x_0$$

$$x_3 = x_2 + h \left.\frac{dx}{dt}\right|_{x_2} = (1 + h\lambda)^3 x_0$$

$$\vdots$$

$$x_k = x_{k-1} + h \left.\frac{dx}{dt}\right|_{x_1} = (1 + h\lambda)^k x_0$$

When $Re(\lambda) < 0$, $x_k \to 0$ as $k \to \infty$ provided $| 1 + h\lambda | < 1$. Representing $\lambda = Re(\lambda) + Im(\lambda)$, the above equation may be rewritten as

$$| (Re(h\lambda) + 1) + Im(h\lambda) | < 1 \tag{2.50}$$

In the $h\lambda$ plane, this corresponds to a circle centered at $(-1, 0)$ with a radius of 1, as shown in the left half plane (corresponding to $Re(\lambda) < 0$ in Figure 2.6(a)). When h satisfies the requirement that $h\lambda$ lies within this region, the resulting solution satisfies the basic requirement of stability: that the asymptotic value of the solution as time tends to ∞ is the same as that for the exact solution.

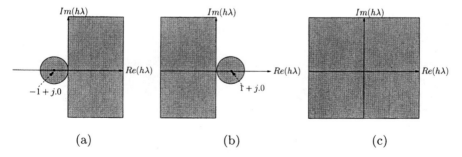

(a)   (b)   (c)

Figure 2.6. Regions of stability for various numerical integration formulæ: (a) the Forward Euler method (b) the Backward Euler method (c) the Trapezoidal rule.

On the other hand, for the case when $Re(\lambda) > 0$, we must have $x_k \to \infty$ as $k \to \infty$. It is easily verified that $(1 + h\lambda)^k \to \infty$ as $k \to \infty$ regardless of what value of h is chosen, implying that the test equation satisfies the stability requirement for this case over the entire right half plane.

Similar analyses for the Backward Euler method and Trapezoidal rule yield regions of stability as shown in Figure 2.6(b) and (c), respectively. Several observations may be made:

- When $Re(\lambda) < 0$, the Backward Euler formula and the Trapezoidal rule are unconditionally stable for the test equation, regardless of the choice of the step size h, while the Forward Euler method satisfies this condition only when h is sufficiently small.

- When $Re(\lambda) > 0$, it is the Forward Euler formula and the Trapezoidal rule that unconditionally obey the requirement that the solution $x_n(t) \to \infty$ as $t \to \infty$, while the Backward Euler method obeys this only when h is sufficiently small.

The significance of the test equation is that the response of many physical system can be represented as a sum of exponentials, so that stability with regard to this equation is necessary. Moreover, since physical systems tend to decay with time, stability in the left half plane is of greater practical interest, and therefore, the Backward Euler method and Trapezoidal rule are "good" methods, while the Forward Euler method is "bad" and is often shunned in practice.

It is important to offer the following caveat: stability is merely a sufficient condition for goodness, and that stability does not imply accuracy. If it did, one would need look no further than the Trapezoidal rule for perfection!

2.5.1   Accuracy and the local truncation error

An alternative interpretation of the Forward Euler and Backward Euler methods views them as truncated Taylor series approximations. Let us consider the solution at the nth time point. Given the solution x_{n-1} at the $(n-1)^{\text{th}}$ time point, we may write the Taylor series approximation of the solution as

$$x_n = x_{n-1} + h \left. \frac{dx}{dt} \right|_{t_{n-1}} + \frac{h^2}{2!} \left. \frac{d^2x}{dt^2} \right|_{t_{n-1}} + \cdots \qquad (2.51)$$

We observe from this equation that the Forward Euler method is merely this approximation, truncated after the first order term. Therefore, we can explicitly say that the error for this approximation is given by the truncated terms. In practice, for a "sufficiently small"[6] value of h, this is dominated by the first truncated term, and we refer to its absolute value as the *local truncation error* (LTE). For the Forward Euler method, this is given by

$$\text{LTE(FE)} = \frac{h^2}{2} \left. \frac{d^2x}{dt^2} \right|_{t_{n-1}} \qquad (2.52)$$

The Backward Euler method can similarly be analyzed, this time with the Taylor series expansion being performed about the point x_n as

$$x_{n-1} = x_n - h \left. \frac{dx}{dt} \right|_{t_n} + \frac{h^2}{2!} \left. \frac{d^2x}{dt^2} \right|_{t_n} + \cdots \qquad (2.53)$$

Again, it is easily seen that the Backward Euler formula arises from the truncation of the terms that are of second order and higher, yielding

$$\text{LTE(BE)} = \frac{h^2}{2} \left. \frac{d^2x}{dt^2} \right|_{t_n} \tag{2.54}$$

The error for the Trapezoidal rule does not emerge quite as simply from the Taylor series approximation, but can be shown to be

$$\text{LTE(TR)} = \frac{h^3}{12} \left. \frac{d^3x}{dt^3} \right|_{x_n} \tag{2.55}$$

Intuitively, the reason why this yields a third order term is that the Trapezoidal rule "averages" the derivative from the Forward Euler and Backward Euler methods, which causes the second-order term to cancel out, leaving the h^3 term as the lowest-order term with a nonzero coefficient.

The expressions for the LTE above can all be seen to be proportional to some power of h. Therefore, if one wanted to control the accuracy, a limit could be set on the allowable LTE in each step, and the step size could be chosen accordingly. For instance, for the Backward Euler method, for a given upper limit ϵ on the LTE, we have

$$\left. \frac{h^2}{2} \frac{d^2x}{dt^2} \right|_{x_n} = \text{LTE(BE)} < \epsilon \tag{2.56}$$

$$\Rightarrow \qquad h < \sqrt{\frac{2\epsilon}{\ddot{x}_n}} \tag{2.57}$$

where $\ddot{x}_n$ is an estimate of the second derivative of x at time point n.

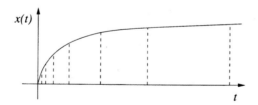

Figure 2.7.   An example to illustrate the utility of a variable time step.

In general, it is possible, and may even be desirable, to have a different value for the time step h at each time point. A specific example is the case of a rising saturated exponential, as shown in Figure 2.7. Initially, when the transition is very quick, a smaller time step is desirable, while in later parts where the value does not change significantly for larger periods of time since the transition is much slower, the time step may be increased without appreciable error. This is automatically detected by Equation (2.57) as $\ddot{x}_n$ changes.

2.5.2   Other numerical integration formulæ

Although this discussion has focused on the Forward Euler and Backward Euler methods and the Trapezoidal rule, a rich variety of numerical integration methods is available. A class of methods that is particularly useful is that of linear multistep formulæ; all of these three methods belong to that class. The order of a linear multistep formula is related to the amount of previous information (values and derivatives at previously computed points) that must be maintained.

Before departing the topic, it is worth pointing to a subset of this class, namely, the set of Gear's formulæ that are "stiffly stable," or in casual parlance, stable for asymptotically decaying systems with steep settling exponentials; the Backward Euler formula, which we know to be stable in the entire left half $h\lambda$ plane, is a first order Gear's formula.

A complete discussion of these techniques is beyond the scope of this book, but the reader is referred to [CL75] for a detailed exposition on the topic in the context of circuit simulation.

2.6 PUTTING IT ALL TOGETHER

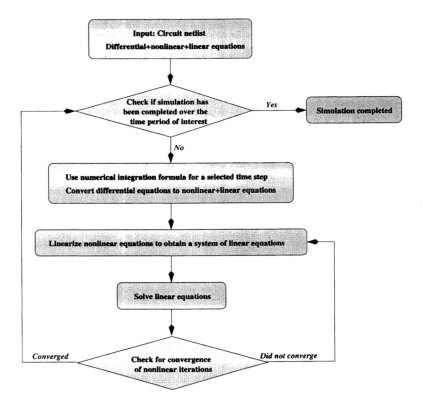

Figure 2.8.   Overall structure of a typical circuit simulator.

The overall structure of a SPICE-like circuit simulator is illustrated in Figure 2.8. The initial system of equations may be a set of linear, nonlinear and differential equations. A time step is chosen and the differential equations are converted to nonlinear equations at each time point using a numerical integration formula. Therefore, at each time step, this results in a system of nonlinear and linear equations that must be solved.

The stamps for the linear equations are entered into the MNA matrix. At each time step, in the inner loop of iterations, the nonlinear equations are linearized at the operating point, as discussed in Section 2.4. The corresponding stamps are entered into the MNA matrix, and the resulting system of linear equations is then solved using techniques similar to those that will be outlined in Section 2.7. Once the nonlinear iterations converge, the outer loop marches on to the next time step, and so on until the entire time period of interest is simulated.

A key issue here is in the generation of the initial conditions for the differential equations, which may not always be provided explicitly, by solving the initial circuit at the steady state. This is referred to as the problem of DC solution, and the reader is referred to [PRV95] for algorithms for this purpose.

2.7 A PRIMER ON SOLVING SYSTEMS OF LINEAR EQUATIONS

From the above discussion, it can be seen that regardless of the type of equation (linear, nonlinear or differential), the circuit simulation procedure transforms it into an approximated linear system of equations that must be solved. Therefore, a vital subproblem in circuit simulation relates to the solution of a system of linear equations, $A\mathbf{x} = \mathbf{b}$, where A is an $n \times n$ matrix, and $\mathbf{x}$ and $\mathbf{b}$ are each $n \times 1$ vectors. We will briefly overview the LU factorization technique, a so-called direct method, and some iterative, or indirect, methods for solving this problem.

2.7.1 LU factorization

The idea of this method is to decompose the matrix A into the product of two $n \times n$ matrices: an lower triangular matrix, L, and an upper triangular matrix, U. To depict this more visually, we write

$$
A = \begin{bmatrix}
a_{11} & a_{12} & \cdots & a_{1n} \\
a_{21} & a_{22} & \cdots & a_{2n} \\
\vdots & \vdots & \ddots & \vdots \\
a_{n1} & a_{n2} & \cdots & a_{nn}
\end{bmatrix}
$$

$$
= \begin{bmatrix}
l_{11} & 0 & \cdots & 0 \\
l_{21} & l_{22} & \cdots & 0 \\
\vdots & \vdots & \ddots & \vdots \\
l_{n1} & l_{n2} & \cdots & l_{nn}
\end{bmatrix} \cdot \begin{bmatrix}
u_{11} & u_{12} & \cdots & u_{1n} \\
0 & u_{22} & \cdots & u_{2n} \\
\vdots & \vdots & \ddots & \vdots \\
0 & 0 & \cdots & u_{nn}
\end{bmatrix} = L \cdot U
$$

If this could be done, the original system could be rewritten as

$$LU\mathbf{x} = \mathbf{b} \tag{2.58}$$

Substituting

$$U\mathbf{x} = \mathbf{y} \tag{2.59}$$

we have

$$L\mathbf{y} = \mathbf{b} \tag{2.60}$$

Since L is a lower triangular matrix, this may be solved easily to find the value of $\mathbf{y}$ using a procedure known as *forward substitution*. Given this value of $\mathbf{y}$, the system in Equation (2.59) may be similarly solved, this time using a method known as *backward substitution*, to find the value of $\mathbf{x}$.

For the specific case where A is symmetric[7] and positive definite[8], it can be shown that one can find an LU factorization so that $L = U^T$, so that we may write $A = LL^T$. This factorization is referred to as the *Cholesky factorization* of A.

The key step is, therefore, in determining how A may be decomposed into the product of L and U. A simple way to understand this is in terms of Gaussian elimination [Gv96], a technique in which we start from the matrix equation $A\mathbf{x} = \mathbf{b}$, and successively perform identical row operations on A and b with the aim of transforming A into an upper triangular matrix. The row transformations successively use *multipliers* m_{ij}, and in each step, subtract m_{ij} times row j from row i, for both the left hand side matrix and the right hand side vector. Any such row transformation preserves the equality between the left and right hand sides. During Gaussian elimination, A eventually becomes an upper triangular matrix U, b is transformed to a vector, $\mathbf{y}$, and the equation takes the form $U\mathbf{x} = \mathbf{y}$.

The careful reader may view this discussion and find the use of notation "sloppy," as the choice of the symbols U and $\mathbf{y}$ in the context of Gaussian elimination above is identical to that used for LU factorization. This is not entirely a mistake, and nor is it accidental! One way to LU-factorize a matrix is to perform Gaussian elimination until the matrix A is transformed into U; at this point, we have Equation (2.59). The matrix L simply corresponds to the multipliers that were used to perform the transformations.

Example Consider the system of equations

$$\begin{bmatrix} 3 & 1 & 1 \\ 1 & 3 & 1 \\ 1 & 1 & 3 \end{bmatrix} \begin{bmatrix} x_1 \\ x_2 \\ x_3 \end{bmatrix} = \begin{bmatrix} 5 \\ 5 \\ 5 \end{bmatrix} \tag{2.61}$$

The sequence of steps through which Gaussian elimination proceeds is listed below; in each case, the multiplier m_{ij} that is used to zero out the $(i,j)^{\text{th}}$ lower triangular element is shown.

$$
\begin{bmatrix} 3 & 1 & 1 \\ 1 & 3 & 1 \\ 1 & 1 & 3 \end{bmatrix} \begin{bmatrix} x_1 \\ x_2 \\ x_3 \end{bmatrix} = \begin{bmatrix} 5 \\ 5 \\ 5 \end{bmatrix} \xRightarrow[m_{31}=1/3]{m_{21}=1/3} \begin{bmatrix} 3 & 1 & 1 \\ 0 & 8/3 & 2/3 \\ 0 & 2/3 & 8/3 \end{bmatrix} \begin{bmatrix} x_1 \\ x_2 \\ x_3 \end{bmatrix} = \begin{bmatrix} 5 \\ 10/3 \\ 10/3 \end{bmatrix}
$$

$$
\xRightarrow{m_{32}=1/4} \begin{bmatrix} 3 & 1 & 1 \\ 0 & 8/3 & 2/3 \\ 0 & 0 & 5/2 \end{bmatrix} \begin{bmatrix} x_1 \\ x_2 \\ x_3 \end{bmatrix} = \begin{bmatrix} 5 \\ 10/3 \\ 5/2 \end{bmatrix} \tag{2.62}
$$

and U is the above upper triangular matrix. Given all of the multipliers, we may simply write the L matrix using these multipliers as

$$
\begin{bmatrix} 1 & 0 & 0 \\ m_{21} & 1 & 0 \\ m_{31} & m_{32} & 1 \end{bmatrix} = \begin{bmatrix} 1 & 0 & 0 \\ 1/3 & 1 & 0 \\ 1/3 & 1/4 & 1 \end{bmatrix} \tag{2.63}
$$

The reader may easily verify that $A = L \cdot U$ for this example   □

To substantiate this more formally, let us now examine the process by which Gaussian elimination transforms $Ax = b$ to $Ux = y$. In each step, the lower triangle of one column i of the matrix is zeroed out through row transformations, by subtracting (m_{ij} times the ith row) from the jth row. Eventually, we can see that

$$
\begin{aligned}
y_1 &= b_1 \\
y_2 &= b_2 - m_{21}y_1 \\
y_3 &= b_3 - m_{31}y_1 - m_{32}y_2 \\
&\vdots \\
y_n &= b_n - m_{n1}y_1 - m_{n2}y_2 - \cdots - m_{n(n-1)}y_{n-1}
\end{aligned}
$$

Moving the negative terms to the left hand side and rewriting the equations in matrix form, we may relate $\mathbf{y}$ and $\mathbf{b}$ through the relation

$$
\begin{bmatrix} 1 & 0 & 0 & \cdots & 0 \\ m_{21} & 1 & 0 & \cdots & 0 \\ m_{31} & m_{32} & 1 & \cdots & 0 \\ \vdots & \vdots & \vdots & \ddots & \vdots \\ m_{n1} & m_{n2} & m_{n3} & \cdots & 1 \end{bmatrix} \mathbf{y} = \mathbf{b} \tag{2.64}
$$

or simply, $L\mathbf{y} = \mathbf{b}$.

We now illustrate the procedures of forward and backward substitution: the terms "forward" and "backward" allude to the order in which the variables are solved for in the system. For the case of forward substitution, since L is lower triangular, one may find the elements of the vector $\mathbf{y}$, starting from the first equation of $L\mathbf{y} = \mathbf{b}$, which directly yields y_1, then proceeding to the second equation, which yields y_2 (since y_1 is known by now), and so on *forward* until we reach the nth equation. Now that $\mathbf{y}$ has been found, solving $U\mathbf{x} = \mathbf{y}$ is similarly easy. Since U is upper triangular, one may start from the last equation to find x_n, then proceed to the second last equation, and so on until the first equation, to find the x_i's in *backward* order, and this explains why the procedure is referred to as backward substitution.

Example For the system of equations (2.61), we solve $L\mathbf{y} = \mathbf{b}$ using forward substitution to obtain the elements of $\mathbf{y}$ as $y_1 = 5, y_2 = 10/3, y_3 = 5/2$. Next, applying backward substitution we find the elements of $\mathbf{x}$ from $U\mathbf{x} = \mathbf{y}$ as $x_3 = 1, x_2 = 1, x_1 = 1$ □

Pivoting

During the process of LU factorization, it is very possible that a diagonal element with a value of zero may be encountered. If so, this would result in a division by zero. This may be avoided by the use of *pivoting*, whereby by exchanging the rows in which equations are placed, or by exchanging the columns that represent any given variable, a nonzero diagonal element, referred to as the pivot, may be obtained, and the processing can continue.

For instance, for the system of equations

$$
\begin{bmatrix} 0 & 2 & 0 \\ 1 & 0 & 0 \\ 0 & 0 & 1 \end{bmatrix} \begin{bmatrix} x_1 \\ x_2 \\ x_3 \end{bmatrix} = \begin{bmatrix} 1 \\ 2 \\ 3 \end{bmatrix}, \tag{2.65}
$$

the first diagonal element is zero. One may exchange the first two equations to obtain

$$
\begin{bmatrix} 1 & 0 & 0 \\ 0 & 2 & 0 \\ 0 & 0 & 1 \end{bmatrix} \begin{bmatrix} x_1 \\ x_2 \\ x_3 \end{bmatrix} = \begin{bmatrix} 2 \\ 1 \\ 3 \end{bmatrix}, \tag{2.66}
$$

which now has a nonzero at the first diagonal element. Alternatively, the first and second variables may be swapped to obtain

$$
\begin{bmatrix} 2 & 0 & 0 \\ 0 & 1 & 0 \\ 0 & 0 & 1 \end{bmatrix} \begin{bmatrix} x_2 \\ x_1 \\ x_3 \end{bmatrix} = \begin{bmatrix} 1 \\ 2 \\ 3 \end{bmatrix}, \tag{2.67}
$$

In general, it is possible to perform a sequence of row and column exchanges to use any arbitrary nonzero in the matrix as a pivot. the element at the (3,3) position in the original matrix may be brought to the first diagonal through a combination of a row exchange and a column exchange to obtain:

$$
\begin{bmatrix} 1 & 0 & 0 \\ 0 & 0 & 1 \\ 0 & 2 & 0 \end{bmatrix} \begin{bmatrix} x_3 \\ x_2 \\ x_1 \end{bmatrix} = \begin{bmatrix} 3 \\ 2 \\ 1 \end{bmatrix}, \tag{2.68}
$$

For efficiency, the choice of pivots may be restricted to the elements of the same column ("column pivoting"), elements of the same row ("row pivoting"), or elements along the diagonal ("diagonal pivoting").

Computational complexity issues

The computational complexity of LU-factorizing an $n \times n$ matrix can be shown to be $O(n^3)$, while that of the forward and backward substitution steps is $O(n^2)$ each. For solving a single system of linear equations, this shows no real gain over a competing method such as Gaussian elimination; in fact, it incurs the extra overhead of storing the L factors. The true advantage of LU-factorization arises when the same left-hand side must be solved repeatedly for a different set of right-hand sides. In the circuit context, this can arise, for example, when the response of a resistive system to a time-varying excitation is required at different times, or when for transient analysis of a system of linear resistors and capacitors under a constant time step h: in each case, the left hand side matrix does not change from one time point to the next, but the right-hand side excitation vector does change. In this case, for k different right hand sides, the $O(n^3)$ expense of LU-factorization is incurred only once, and given these LU factors, forward and backward substitution are repeatedly carried out for the k excitations, yielding a complexity of $O(n^3 + kn^2)$, as against $O(kn^3)$ for Gaussian elimination.

For real circuits, additional savings may be obtained. Recall that the stamps for most of the elements discussed in Section 2.2.1 yielded a constant number of entries in the left hand side matrix; systems of mutual inductances are a notable exception. Therefore, even for an MNA system with n variables, the number of nonzero entries in the MNA matrix is $O(n)$, which is much less than the size of the matrix, n^2. Such a matrix where the number of nonzero entries is small is referred to as a *sparse* matrix, and smart techniques may be used to reduce the cost of LU-factorization. Empirically, for typical circuit matrices, it has been observed that the cost of LU factorization ranges from "$O(n^{1.2})$" to "$O(n^{1.7})$"[9].

2.7.2   Indirect methods

An indirect method for solving linear equations begins with a guess solution and updates the solution through a set of iterations. In some cases, the iterations will converge regardless of the initial guess, while in others, the iterations will converge under some conditions but not under others. We will examine some of the related issues in this section.

To begin with, let us consider a system of linear equations $A\mathbf{x} = \mathbf{b}$, or more completely,

$$
\begin{bmatrix}
a_{11} & a_{12} & \cdots & a_{1n} \\
a_{21} & a_{22} & \cdots & a_{2n} \\
\vdots & \vdots & \ddots & \vdots \\
a_{n1} & a_{n2} & \cdots & a_{nn}
\end{bmatrix}
\begin{bmatrix}
x_1 \\
x_2 \\
\cdots \\
x_n
\end{bmatrix}
=
\begin{bmatrix}
b_1 \\
b_2 \\
\cdots \\
b_n
\end{bmatrix}
\tag{2.69}
$$

and an initial guess $\mathbf{x}^0 = [x_1^0 x_2^0 \cdots x_n^0]^T$.

In the $(k+1)^{\text{th}}$ iteration, one could use the ith equation to update the value of $\mathbf{x}$ as follows:

$$x_1^{k+1} = \frac{b_1 - a_{12}x_2^k + a_{13}x_3^k + a_{14}x_4^k \cdots + a_{1n}x_n^k}{a_{11}}$$

$$x_2^{k+1} = \frac{b_2 - a_{21}x_1^k + a_{23}x_3^k + a_{24}x_4^k \cdots + a_{2n}x_n^k}{a_{22}}$$

$$\vdots$$

$$x_n^{k+1} = \frac{b_n - a_{n1}x_1^k + a_{n2}x_2^k + a_{n3}x_3^k \cdots + a_{n(n-1)}x_{n-1}^k}{a_{nn}} \tag{2.70}$$

Alternatively, observing that if these computations are carried out sequentially, at the time when x_i^{k+1} is to be computed, x_j^{k+1} is known $\forall j < i$, we may use the following update formula instead:

$$x_1^{k+1} = \frac{b_1 - a_{12}x_2^k + a_{13}x_3^k + a_{14}x_4^k \cdots + a_{1n}x_n^k}{a_{11}}$$

$$x_2^{k+1} = \frac{b_2 - a_{21}x_1^{k+1} + a_{23}x_3^k + a_{24}x_4^k \cdots + a_{2n}x_n^k}{a_{22}}$$

$$\vdots$$

$$x_n^{k+1} = \frac{b_n - a_{n1}x_1^{k+1} + a_{n2}x_2^{k+1} + a_{n3}x_3^{k+1} \cdots + a_{n(n-1)}x_{n-1}^{k+1}}{a_{nn}} \tag{2.71}$$

The update formula in Equation (2.70) is referred to as the Gauss-Jacobi method, while the formula in Equation (2.71) is called the Gauss-Seidel method. The former has the advantage over the latter of being easier to parallelize, since each variable in the $(k+1)^{\text{th}}$ iteration may be simultaneously updated; this is not possible for the latter due to the sequential dependencies between the update formulæ for x_i^{k+1} and $x_j^{k+1} \forall j > i$.

While there are various ways of characterizing the requirements for convergence, it can be shown that if the matrix A is diagonally dominant[10], both the Gauss-Seidel and Gauss-Jacobi methods converge unconditionally, regardless of the initial guess. This is a particularly useful fact, since there are practical circuit topologies for which the nodal or MNA matrix is diagonally dominant.

The successive overrelaxation (SOR) method is an alternative update formula that sets the next update value for a variable to be

$$x_i^{k+1} = (1 - \omega)x_i^k + \omega x_i^{k+1}(GS) \tag{2.72}$$

where $x_i^{k+1}(GS)$ is the updated value using the Gauss-Seidel update formula from Equation 2.71. This formula uses an extrapolation that finds the update as a weighted sum of the previous iterate and the Gauss-Seidel update. It is clear that for $\omega = 1$, this method is identical to the Gauss-Seidel method. Moreover, it has been proven that if $\omega \notin (0, 2)$, the procedure will not converge.

More sophisticated methods do exist, such as the conjugate gradient method or the GMRES (generalized minimum residual) method, and for a detailed explanation, the reader is referred to [Gv96, SS86].

A technique that is often used to hasten the convergence of an iterative method involves the use of *preconditioners*. Given a system of n equations in n variables, $A\mathbf{x} = \mathbf{b}$, and an $n \times n$ matrix P that acts as a preconditioner, one may premultiply both sides of the matrix equation by P to obtain

$$PA\mathbf{x} = P\mathbf{b} \tag{2.73}$$

If P were to be exactly A^{-1}, then the method would conclude in a single iteration; this however, involves the substantial computational effort of finding A^{-1}. Instead, if P were to be a "good" approximation to A^{-1}, then PA would be "close" to the identity matrix, and the iterations would converge relatively soon. The notion of preconditioning essentially involves the choice of a suitable P that is both easy to compute and satisfies the requirement of being a "good" approximation to A^{-1}.

For a detailed discussion on preconditioning, the reader is referred to [Gv96].

2.8 SUMMARY

This chapter has presented an overview of techniques used for circuit simulation. For elements represented by linear device equations, the MNA formulation is employed, and we have seen how it may be constructed by inspection of the circuit. Nonlinear elements are iteratively linearized at a guess solution and folded into this formulation, and elements described by differential equations are numerically integrated, so that the system is solved at successive time points. This represents the slowest and practically most exact way of solving a circuit. In the succeeding chapters, we will next see methods that can be used for faster simulation.

Notes

1. The reader is invited to verify in the subsequent discussion that this choice merely provides a reference direction, and does not affect the correctness or validity of the circuit equations.

2. The rank of a matrix is the dimension of its largest square submatrix that has a nonzero determinant (i.e., is nonsingular).

3. In this context, this means that if G were to be converted to an undirected graph by removing the edge directions, then there would be a path between any pair of nodes $(i, j) \in V$.

4. Pedantically, this should be referred to as an affine approximation, but we will refer to it through the (strictly speaking, incorrect) common usage that refers to it as linear.

5. Note that if a nonlinear capacitor were used instead, then the fundamental quantity that describes it would be the charge q on the capacitor. The characteristic equation would then be

$$\frac{dq}{dt} = i, \quad q = C(v)v \tag{2.74}$$

The capacitance is denoted here as $C(v)$ to emphasize that it is a function of the voltage. It is particularly important to observe this since many of the capacitances in MOS circuits are indeed nonlinear, and the blind use of numerical integration formulæ based on Equation (2.42) instead of Equation (2.74) can lead to significant errors.

6. This sufficiently small value should lie within the radius of convergence of the Taylor series. Note that this is purely related to accuracy, and should not be confused with stability issues.

7. A symmetric matrix is one for which $a_{ij} = a_{ji} \forall i, j$. In other words, $A = A^T$.

8. A matrix A is positive definite [positive semidefinite] if for every nonzero vector $\mathbf{x}$, the scalar $\mathbf{x}^T A \mathbf{x} > 0 \ [\geq 0]$. An alternative characterization of a positive definite [semidefinite] matrix is that its eigenvalues are all larger than [larger than or equal to] zero.

9. The quotes refer to the fact that this is not a worst-case complexity, and is merely found empirically over typical circuits.

10. A matrix A is said to be diagonally dominant if $|a_{ii}| \geq \sum_{j=1 \text{ to } n, j \neq i} |a_{ij}|$. The matrix is said to be strictly diagonally dominant if the inequality is strict.

3   FREQUENCY-DOMAIN ANALYSIS OF LINEAR SYSTEMS

3.1   INTRODUCTION

A typical digital circuit consists of a number of transistors, often organized into gates, that drive interconnect wires. The transistors may be modeled with the aid of nonlinear resistances and capacitances, while the behavior of the wires is represented by resistors, capacitor and inductors. As described in Chapter 2, calculating all of the currents and voltages in such a circuit over a specified period of time requires its solution at numerous time steps, at each of which the nonlinear equations are iteratively solved. Under circumstances where much more limited accuracy is desired, or when only some summary attributes of the waveform are required, such a time-domain analysis method is far too elaborate and time-consuming, and faster *timing analysis* techniques are employed.

The chief characteristic of digital circuits that permits simplified circuit simulation is that they can be assumed to be decomposed into stages of logic gates, and each of these stages can reasonably be considered independently of each other. In this chapter, we will focus on techniques that can be used to find the delay of a *linear system*, such as an interconnect net. In Chapter 4, we will utilize these results to first compute the delay of a stage of logic that consists of an interconnect driven by a logic gate, and then use this as a building block to compute the delay of an entire combinational circuit in Chapter 5.

Before proceeding, it is appropriate to focus on two attributes of a waveform, pictorially described in Figure 3.1, that are of great interest during design:

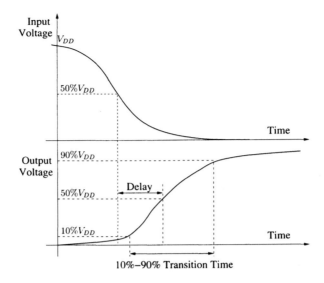

Figure 3.1.   Pictorial depiction of the 50% delay and transition time.

The 50% delay (often referred to simply as the delay) of a waveform relates to the amount of time from the instant when a stimulus is applied to when its response asserts itself. In the context of a digital circuit, it is defined as the interval that elapses between the time when the input waveform crosses a specified threshold, and when the output waveform crosses a given threshold. Although these two thresholds could, in principle, be different from each other, they are most commonly set to be at the halfway point, also known as the 50% point, of the waveform transition.

The transition time of a waveform is related to the "slope" of the waveform, and is typically defined as the time in which the waveform goes from $x\%$ to $y\%$ of its final value. Since most such transitions involve exponentials, the most commonly used measures correspond to the 10%-90% transition time and the 20%-80% transition time. However, if the signal is modeled as a piecewise linear function (such as a saturated ramp), a 0% to 100% transition time is meaningful.

Although finding the most exact values of parameters such as the delay or the transition time of a waveform requires rigorous circuit simulation, quicker and more approximate estimates may be obtained using timing analysis techniques.

A notable fact is that it is indeed possible, under this definition of the 50% delay, to achieve a negative delay! An extremely slow input waveform applied to the input of a logic stage, as shown in Figure 3.2, may actually result in the output rising at a faster rate, as a result of which the difference between the 50% crossing points of the output and the input is negative. This is not

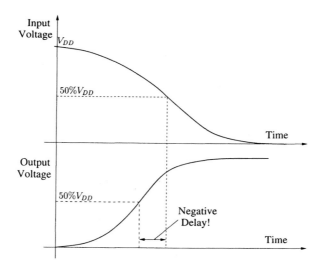

Figure 3.2.   When the input waveform has a particularly slow transition and the output rises relatively quickly, so that its 50% crossing time is earlier than that of the input, the resulting delay can be negative.

a nonphysical effect, but merely a consequence of the artificial nature of the choice of the 50% point as the delay threshold. Such a scenario may occur when, for example, a gate with low width transistors is used to drive a large output load[1].

Let us now consider the problem of computing these values for the case of an an inverter driving an interconnect wire. During a transition, the inverter can be represented by a set of nonlinear resistors and capacitors, while the interconnect can be represented by a set of linear resistors, capacitors and possibly, inductors (we will consider these models in further detail in Section 3.2). For the time being, let us ignore the nonlinearities and assume all elements to be perfectly linear; we will consider the consequences of this assumption and revisit it later in Chapter 4. For such an RLC system, one can think of the response in terms of a set of *time constants* that dictate the transition, and these time constants depend on the RLC values. If one could estimate the dominant time constants of such a system, it would be a useful aid in calculating the delay of the system. In the succeeding discussion, we will consider first the *Elmore delay metric*, which attempts to identify the single dominant time constant of a system, and then the *asymptotic waveform evaluation* (AWE) method and other higher order modeling techniques, which generalizes this to find multiple time constants.

To understand the notion of a dominant time constant, let us consider the example circuit shown in Figure 3.3. The exact response at node 5 to a unit step excitation at the input is given in terms of the frequency variable s as

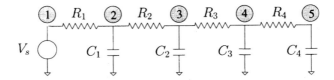

Figure 3.3.   An example of an RC line.

follows:

$$V_5(s) = \frac{1}{s(1 + 10s + 15s^2 + 7s^3 + s^4)} \tag{3.1}$$

$$= \frac{1}{s} - \frac{1.24}{s + 0.12} + \frac{0.33}{s + 1} - \frac{0.12}{s + 2.347} + \frac{0.028}{s + 3.53} \tag{3.2}$$

where the latter expression corresponds to a partial fraction expansion. In the time domain, this corresponds to the following sum of exponentials

$$V_5(t) = 1 - 1.24e^{-0.12t} + 0.33e^{-t} - 0.12e^{-2.347t} + 0.028e^{-3.53t} \tag{3.3}$$

The transient here is dominated by the time constant of the first exponential term, which corresponds to the first few exponential terms, and the pole at $s = -0.12$ is referred to as a *dominant pole*[2]. While such a distinction is qualitative rather than quantitative, if one were to approximate the exact system by a lower-order system that captures the dominant pole(s) of the original system, a very good approximation to the waveform could be obtained. Figure 3.4 show the results of applying the AWE technique (to be discussed in Section 3.5) to the above system and it is apparent that even a first or second order approximation is adequate to capture the most important characteristics of this waveform.

This chapter will discuss techniques for the reduced order modeling of linear systems, starting with the asymptotic waveform evaluation method, and then progressing into Krylov subspace-based methods. For a more in-depth description, the reader is referred to [CN94, CPO02].

3.2 INTERCONNECT MODELING

Interconnect wires may be modeled at various levels of abstraction, ranging from relatively simple models to those that arise from a full 3-dimensional extraction. We will outline a set of low-frequency models from the former category here, which will consider resistance and capacitance, but not inductance, and for inductance extraction approaches, the reader is referred to, for example, [BP01, GBW+01]. The analysis methods described later in this chapter, however, are valid for RLC systems.

Given a wire of length l and width w, its resistance R may be modeled as

$$R = \rho_s \frac{l}{w} \tag{3.4}$$

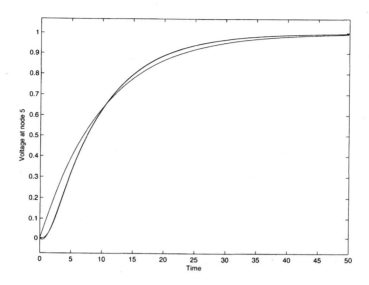

Figure 3.4. A comparison of the first and second order AWE waveforms with the exact step response voltage waveform at node 5 in Figure 3.3. The second order response nearly coincides with the exact response.

where ρ_s is the sheet resistance, which is the ratio of the resistivity of the material and the thickness of the wire. A simple model for the line-to-ground wire capacitance is given by

$$C_g = \beta lw + \gamma l \tag{3.5}$$

where β is the area capacitance and γ is the fringing capacitance. The line-to-line capacitance for between two wires that have a coupling length of l_c and a separation of d may be characterized by

$$C_c = \chi \frac{l_c}{d^\alpha} \tag{3.6}$$

where χ is the coupling capacitance per unit length, and α is an empirical constant whose value is between 1 and 2 (a frequently-used value for α is 1.34).

Other, more complex, models for the interconnect capacitance such as those proposed by Sakurai [ST83] and Chern [CHA+92], are frequently employed.

An on-chip wire with a ground plane is most accurately represented as a distributed transmission line, with infinitesimal elements of size dx, with a resistance of $r \cdot dx$ and a capacitance to ground of $c \cdot dx$, where r and c are the per unit resistance and capacitance, respectively. However, practically, this may be represented by discrete, noninfinitesimal segments with lengths of the order of a hundred microns, each associated with lumped elements in the form

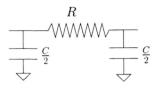

Figure 3.5.   A π model for a segment of an interconnect wire.

of π models, illustrated in Figure 3.5. The total resistance R of the segment is connected between the two ends of the segment, and the total capacitance to ground, C, is divided into two parts, one placed at each end of the segment. This basic π model can be used as a building block to model larger lines: for example, for a 1000μm line with a segment length of 100μm, ten such π models are cascaded together to represent the line. For coupled lines, the pi model could be extended as follows. The coupling capacitance associated with each segment could be divided into two components, one at each end of the segment, and this could be connected between the segment under consideration and the segment that it couples to. An example of an RC circuit that results from this is shown in Figure 3.8.

3.3 TYPICAL INTERCONNECT STRUCTURES

We will define a set of basic structures that are commonly encountered in analyzing interconnect circuits. An *RC tree* is a tree-structured connection of resistors with the requirement that any capacitance connected to a node of the tree must have its other terminal connected to ground, and that no resistor is connected to ground. A *distributed RC tree* is a tree structure where each branch corresponds to a distributed resistance between nonground nodes, and a distributed capacitance to ground; this differs from an RC tree, which contains lumped resistors for each wire segment.

The tree-structured nature of the resistances naturally leads to two properties: firstly, resistive loops are forbidden, and secondly, if one of the nonground nodes is considered to be an input, then there is a unique path from the input to any nonground node of the tree. An example of an RC tree is shown in Figure 3.6. The input node of the network is the node n_0, and is driven by the excitation, V_s.

A specific case of an RC tree is an *RC line*, which corresponds to an RC tree where all nodes are connected to two resistors, except the nodes at the far ends that are each connected to a single resistor. Coarsely speaking, this is an RC tree without branches, and an instance of an RC line is illustrated in Figure 3.3.

An *RC mesh* is similar to an RC tree, except that resistor loops are permitted in RC meshes; an example of an RC mesh is shown in Figure 3.7. Clearly, since

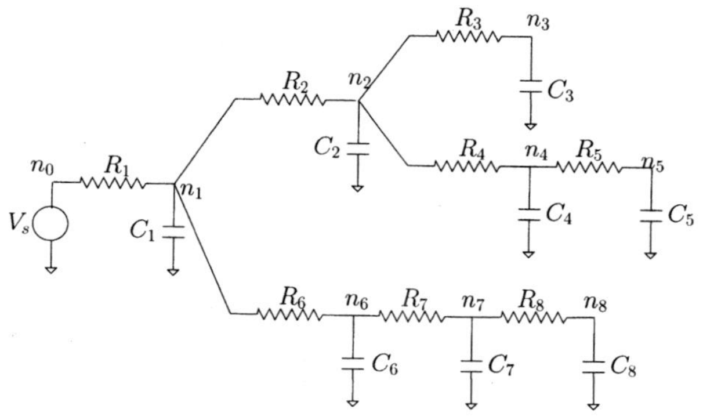

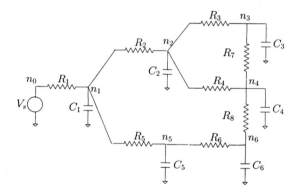

Figure 3.6.   An example of an RC tree.

any RC tree is (trivially) an RC mesh, the set of RC meshes is a superset of the set of RC trees.

Figure 3.7.   An example of an RC mesh.

An *RLC tree* and an *RLC mesh* may be defined in an analogous manner to the RC tree and the RC mesh, respectively, when each resistance is replaced by a resistor in series with a self-inductor.

It is also common to encounter *coupled RC lines*, which can be thought of as a set of RC lines that are connected by floating (i.e., non-grounded) capacitances, as illustrated in Figure 3.8. A coupled RLC line consists of a set of RLC lines that are coupled through mutual inductances and/or coupling capacitors. One may similarly define coupled RC trees, coupled RC meshes, coupled RLC trees and coupled RLC meshes.

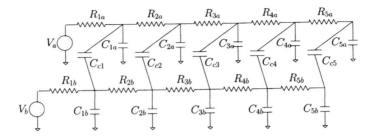

Figure 3.8.   An example of coupled RC lines.

3.4   THE ELMORE DELAY METRIC

The Elmore delay provides a useful technique for estimating the delay of circuits whose response is well-captured by a dominant time constant. As will be shown in Section 3.5, this metric corresponds to a first order AWE approximation of a circuit, and in cases where multiple time constants affect the behavior of the system, its accuracy may be limited. Nevertheless, there is a wide range of applications for which it is of great utility.

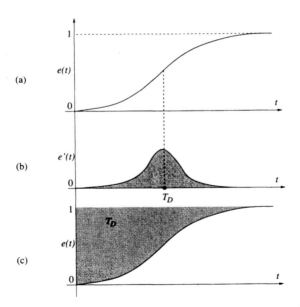

Figure 3.9.   Interpretations of the Elmore delay metric. The step response $e(t)$ at some node in a circuit is shown in (a) and its derivative in (b). The time coordinate of the centroid of $e'(t)$, is the Elmore delay, T_D. The shaded area above the step response in (c) is also equal to T_D.

The Elmore metric can be understood in terms of a simple and intuitively appealing description. Consider, for example, an RC mesh that is initially rested, i.e., the initial voltage at any node is zero. If a unit step excitation is applied to this structure, then the voltage at each node will monotonically increase from 0 to 1. Figure 3.9(a) illustrates such a waveform, $e(t)$, at some non-source node of this RC network. We will refer to the time derivative of this waveform as $e'(t)$.

In [Elm48], Elmore made the observation that the time coordinate of the center of area of the region under the curve $e'(t)$ (illustrated by the shaded region in Figure 3.9(b)) would serve as a reasonable estimate of the delay. In other words, an approximation to the 50% delay, T_D, may be expressed as[3]

$$T_D = \int_0^\infty te'(t)dt, \tag{3.7}$$

From Figure 3.9(a), such a metric can be seen to have great intuitive appeal.

From elementary linear system theory, we observe that since $e(t)$ is the step response of the rested system, $e'(t)$ must be the response to an excitation that is the derivative of the step response, namely, the unit impulse. Hence, Equation (3.7) also has the interpretation of being the *first time moment of the impulse response.*

An interesting observation is that the quantity T_D is also the area above the step response, as shown in Figure 3.9(c). This can be seen by performing integration by parts, as shown below:

$$\begin{aligned}
T_D &= \int_0^\infty te'(t)dt \\
&= \int_0^\infty t\frac{d}{dt}(e(t) - 1) \\
&= t\left[e(t) - 1\right]|_0^\infty - \int_0^\infty \left[e(t) - 1\right] dt \\
&= \int_0^\infty \left[1 - e(t)\right] dt \tag{3.8}
\end{aligned}$$

Here, we make use of the fact that $\lim_{t \to \infty} t\left[1 - e(t)\right] = 0$ since $e(t) \to 1$ as $t \to \infty$, and the latter term decreases exponentially with t while the former term only increases linearly.

3.4.1   An expression for the Elmore delay through RC networks

Although the above expressions provide formal definitions of the Elmore delay metric, they are rather cumbersome and difficult to use. In this section, we will derive a more usable form for the Elmore delays in a general RC tree or mesh structure (with no floating capacitors), and derive a closed-form expression for the Elmore delays in RC trees and RC meshes.

For a given RC mesh with n grounded capacitors, the MNA matrix of conductances defined in Section 2.2 has the following form

$$G = \begin{bmatrix} g_{1,1} & -g_{1,2} & \cdots & -g_{1,n} \\ -g_{2,1} & g_{2,2} & \cdots & -g_{2,n} \\ \vdots & \vdots & \ddots & \vdots \\ -g_{n,1} & -g_{n,2} & \cdots & g_{n,n} \end{bmatrix}, \tag{3.9}$$

where $g_{i,j} = g_{j,i}$ is the branch conductance between nodes i and j, and $g_{i,i}$ is the sum of all branch conductances connected to node i. We define the capacitance matrix C and the corresponding capacitance vector $\mathbf{C}$ as

$$C = \begin{bmatrix} C_1 & 0 & 0 & \cdots & 0 \\ 0 & C_2 & 0 & \cdots & 0 \\ 0 & 0 & C_3 & \cdots & 0 \\ \vdots & \vdots & \vdots & \ddots & \vdots \\ 0 & 0 & 0 & \cdots & C_n \end{bmatrix} \qquad \mathbf{C} = \begin{bmatrix} C_1 \\ C_2 \\ C_3 \\ \vdots \\ C_n \end{bmatrix}, \tag{3.10}$$

where C_i is the value of the grounded capacitance at node i.

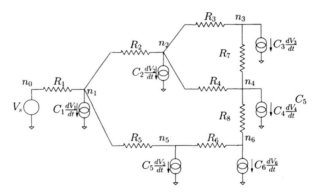

Figure 3.10. An example to illustrate the replacement of capacitors by current sources in the circuit in Figure 3.7.

A very widely used result states that the Elmore delay, T_{Di}, to the ith node of an RC network is given by the ith element of the $n \times 1$ vector

$$\mathbf{T_D} = G^{-1}\mathbf{C} \tag{3.11}$$

The proof of this statement is as follows [Wya87]. The current flowing through all capacitors can be abstracted as dependent current sources that inject the current

$$-C_i \frac{dV_i}{dt} \tag{3.12}$$

into node i, where V_i is the voltage at node i. This is illustrated by the example in Figure 3.10.

This modified network now consists entirely of resistors and current sources. Taking the input node, which is at voltage of 1 after $t = 0$, as the datum, its nodal equations for $t > 0$ may be written as

$$
\begin{bmatrix}
g_{1,1} & \cdots & -g_{1,n} \\
-g_{2,1} & \cdots & -g_{2,n} \\
\vdots & \ddots & \vdots \\
-g_{n,1} & \cdots & g_{n,n}
\end{bmatrix}
\begin{bmatrix}
V_1 - 1 \\
V_2 - 1 \\
\vdots \\
V_n - 1
\end{bmatrix}
= -
\begin{bmatrix}
C_1 & 0 & \cdots & 0 \\
0 & C_2 & \cdots & 0 \\
\vdots & \vdots & \ddots & \vdots \\
0 & 0 & \cdots & C_n
\end{bmatrix}
\begin{bmatrix}
\dot{V}_1 \\
\dot{V}_2 \\
\vdots \\
\dot{V}_n
\end{bmatrix}
$$

$$(3.13)$$

In other words,

$$G(\mathbf{V} - \mathbf{1}) = -C\dot{\mathbf{V}}$$

i.e., $\quad \mathbf{1} - \mathbf{V} = G^{-1}C\dot{\mathbf{V}}$

$$(3.14)$$

where the $n \times 1$ vector, $\mathbf{1}$, has all entries equal to 1.

Since, from Equation (3.8), the Elmore delay at node i is given by $T_{Di} = \int_0^\infty [1 - V_i(t)]\,dt$, we can write

$$
\begin{aligned}
\mathbf{T_D} &= \int_0^\infty [\mathbf{1} - \mathbf{V}(t)]\,dt \\
&= \int_0^\infty G^{-1}C\dot{\mathbf{V}}\,dt \\
&= G^{-1}C\,[\mathbf{V}(\infty) - \mathbf{V}(0)] \\
&= G^{-1}C\mathbf{1} \\
&= G^{-1}\mathbf{C}
\end{aligned}
$$

$$(3.15)$$

3.4.2   A closed-form Elmore delay expression for RC trees

In general, inverting G is a tedious and computationally expensive procedure. However, in the case of an RC tree, G^{-1} admits a closed form, and the Elmore delay at node n_i can be written down by inspection as follows. Let P_i be the unique path from the input node n_0 to some internal node n_i of the tree, and for any two nodes n_i and n_j in the tree, we will denote $P_{ij} = P_i \cap P_j$ as the portion of the path between n_0 and n_i that is common to the path between n_0 and n_j. The set of the resistances in the path P_{ij} are referred to as the *upstream resistances* for node j.

The Elmore delay to node n_i in the RC tree is then given by the expression

$$T_{Di} = \sum_{j=0}^{n} C_j \sum_{k \in upstream(j)} R_k \qquad (3.16)$$

By rearranging the terms, it can be shown that this expression may also be rewritten as

$$T_{Di} = \sum_{j \in P_i}^{n} R_j \sum_{k \in downstream(j)} C_k \tag{3.17}$$

where the downstream capacitance at j is the sum of all capacitances at any node k such that the unique path in the tree from j to the root must pass through k, i.e., $P_k \subseteq P_j$.

Example: Consider the circuit shown in Figure 3.6. The node n_0 is the input node. The procedure described above may be used to compute the Elmore delay to various nodes in the circuit. For example, using the upstream resistance formula, we can compute

$$
\begin{aligned}
T_D(n_7) &= R_1C_1 + R_1C_2 + R_1C_3 + R_1C_4 + R_1C_5 + (R_1 + R_6)C_6 \\
&\quad + (R_1 + R_6 + R_7)C_7 + (R_1 + R_6 + R_7)C_8. \tag{3.18} \\
T_D(n_5) &= R_1C_1 + (R_1 + R_2)C_2 + (R_1 + R_2)C_3 \\
&\quad + (R_1 + R_2 + R_4)C_4 + (R_1 + R_2 + R_4 + R_5)C_5 \\
&\quad + R_1C_6 + R_1C_7 + R_1C_8. \tag{3.19}
\end{aligned}
$$

The downstream capacitance formula may also be used to write these terms as

$$
\begin{aligned}
T_D(n_7) &= R_1(C_1 + C_2 + C_3 + C_4 + C_5 + C_6 + C_7 + C_8) \\
&\quad + R_6(C_6 + C_7 + C_8) + R_7(C_7 + C8) \tag{3.20} \\
T_D(n_5) &= R_1(C_1 + C_2 + C_3 + C_4 + C_5 + C_6 + C_7 + C_8) \\
&\quad + R_2(C_2 + C_3 + C_4 + C_5) + R_4(C_4 + C_5) + R_5C_5 \tag{3.21}
\end{aligned}
$$

The reader is invited to verify that the expressions are indeed identical.

Several techniques [LM84, Wya85, CS89, CK89] have been suggested for a fast computational approximation to the Elmore delay of a general RC mesh.

3.5   ASYMPTOTIC WAVEFORM EVALUATION

The Elmore delay metric turns out to be the simplest first order expression of the more general idea of model order reduction that was alluded to earlier, in which a higher order system is approximated by a lower order system that captures the most important characteristics, such as the dominant poles, of the original system. The technique of Asympotic Waveform Evaluation (AWE) [PR90] provides a systematic technique for performing such a reduction based on the moments of the transfer function for a system. In this section, we will first define the moments of a transfer function and then develop a systematic method for computing them. We will then present the engine of AWE, a computationally efficient technique for performing model order reduction, and finally, discuss the limitations of the method.

3.5.1   Moments of a transfer function

Consider a single-input single-output linear system whose transfer function[4] is represented by the function $f(t)$. The Laplace transform of this transfer function is given by

$$F(s) = \int_0^\infty f(t)e^{-st}dt \qquad (3.22)$$

A MacLaurin series expansion (i.e., a Taylor series approximation about $s = 0$) of the exponential results in the expression

$$
\begin{aligned}
F(s) &= \int_0^\infty f(t)\left[1 - st + \frac{s^2t^2}{2!} - \frac{s^3t^3}{3!} + \cdots\right]dt \\
&= \int_0^\infty f(t)dt - s\int_0^\infty tf(t)dt + s^2\int_0^\infty \frac{t^2}{2!}f(t)dt - s^3\int_0^\infty \frac{t^3}{3!}f(t)dt + \cdots \\
&= m_0 + m_1s + m_2s^2 + m_3s^3 + \cdots \qquad (3.23)
\end{aligned}
$$

where

$$m_i = (-1)^i \int_0^\infty \frac{t^i}{i!}f(t)dt \qquad (3.24)$$

corresponds to the ith moment of the transfer function $f(t)$.

Once the moments of the transfer function are calculated, a technique known as the Padé approximation[5] may be applied to easily compute a reduced order model for the system using a process known as *moment matching*. A Padé approximant is merely a lower order transfer function, and it is characterized by its order, denoted $[L/M]$, where L is the order of its numerator polynomial and M the order of the denominator polynomial.

To convey the essential idea of moment matching, let us consider the following example that matches the moments of an arbitrary transfer function to a reduced order model that corresponds to a first-order transfer function. In essence, what we try to do is to assert that

$$m_0 + m_1s + m_2s^2 + m_3s^3 + \cdots \equiv \frac{a_0}{1 + b_1s} \qquad (3.25)$$

Here, we assume that the m_i's are all known, and that the unknowns correspond to a_0 and b_1. Cross-multiplying the denominator of the right-hand side with the left-hand side, we have

$$(m_0 + m_1s + m_2s^2 + m_3s^3 + \cdots)(1 + b_1s) \equiv a_0 \qquad (3.26)$$

For the two sides to "match," the coefficient of each power of s must be equal. In other words, if we consider each power of s separately, we have:

$$
\begin{aligned}
s^0 &: \quad m_0 &= a_0 \qquad (3.27) \\
s^1 &: \quad m_1 + m_0b_1 &= 0 \qquad (3.28)
\end{aligned}
$$

While this process could continue indefinitely for arbitrary powers of s, since we have only two unknowns, any further matching could result in an overdetermined system with no solution. Therefore, we choose not to try to match any further powers of s beyond the first, and obtain

$$a_0 = m_0$$
$$b_1 = -m_1/m_0 \qquad (3.29)$$

Relation to the Elmore delay metric. The Elmore delay corresponds to an AWE approximation to a first-order system. To see this, consider any RC structure with no resistances to ground, such as an RC tree or an RC mesh excited by a unit step input. In the steady state the capacitors carry no current, and therefore, the voltage $V(t)$ at any nonground node reaches unity. If the moments of the transfer function to a node are given by Equation (3.23), the step response can be obtained by multiplying this by $1/s$, which is the Laplace transform of the unit step excitation. We can apply the final value theorem [Kuo93] as follows. We know that for a step input, the steady state voltage at any node of an RC tree after the transient has settled must be given by

$$\lim_{t \to \infty} V(t) = 1$$

Since

$$\lim_{t \to \infty} V(t) = \lim_{s \to 0} sV(s) = \lim_{s \to 0} s \cdot \frac{1}{s} \cdot (m_0 + m_1 s + m_2 s^2 + \cdots) = m_0, \quad (3.30)$$

this implies that $m_0 = 1$ for the voltage step response at any node in an RC tree. In conjunction with Equation (3.29), this means that the [0/1] Padé approximation for the voltage step response at any node of an RC tree is

$$V(s) = \frac{1}{s} \cdot \frac{1}{1 - m_1 s} \qquad (3.31)$$
$$\text{i.e., } V(t) = 1 - e^{t/m_1} \qquad (3.32)$$

In other words, the Elmore delay at a node in an RC structure with no grounded resistors is the negation of the time constant of its step response transient when it is approximated as a first order system. Specifically:

- The 50% delay is given by $-m_1 \log 2$.

- The 10%-90% transition time is given by $-2.2m_1$.

3.5.2 *Efficient moment computation*

From Section 2.2.1, we can see that an RLC interconnect system may be represented by an MNA matrix that can be constructed by using an aggregation of element stamps. If we represent resistors and capacitors using the stamps of Equations (2.15) and (2.19), respectively, and inductors as Type 2 elements (as defined in Section 2.2), with stamps given by Equation (2.20), then, using the MNA formulation from Chapter 2, an RLC system may be represented as:

$$(G + sC)\mathbf{X} = \mathbf{E} \tag{3.33}$$

where G and C are constant matrices whose entries depend on the values of the RLC elements, $\mathbf{X}$ is the vector of unknowns, consisting of node voltages and currents in Type 2 elements such as inductors and voltage sources, and $\mathbf{E}$ is the excitation vector.

If we represent $\mathbf{X}$ in terms of its moments, we have:

$$\mathbf{X} = \mathbf{m}_0 + \mathbf{m}_1 s + \mathbf{m}_2 s^2 + \mathbf{m}_3 s^3 + \cdots \tag{3.34}$$

Note that the jth entry of vector $\mathbf{m}_i$ is simply the ith moment of the jth variable.

We will use this to find the impulse response of the system, i.e., the response of the system to a $\delta(t)$ excitation. Under this condition, we have $\mathbf{E} = \mathbf{E}_0$, a constant vector in the s domain. Substituting these two into Equation (3.33),

$$(G + sC)(\mathbf{m}_0 + \mathbf{m}_1 s + \mathbf{m}_2 s^2 + \cdots) = \mathbf{E}_0 \tag{3.35}$$

Matching the powers of s on either side of the equation, we obtain the following relations:

$$G\mathbf{m}_0 = \mathbf{E}_0 \tag{3.36}$$
$$G\mathbf{m}_i = -C\mathbf{m}_{i-1} \ \forall \ i \geq 1 \tag{3.37}$$

The above set of equations has a very useful physical interpretation, related to the original system that is described by Equation (3.33).

- The solution of Equation (3.36) is identical to that of the original system when an impulse excitation is applied, and s is set to zero. In other words, it is the response of the original circuit to a unit impulse at DC, i.e., with all capacitors open-circuited and all inductors short-circuited.

- Equation (3.37) corresponds to Equation (3.33) if s is set to zero, the original excitation is set to zero, and a new excitation of $-C\mathbf{m}_{i-1}$ is applied instead. This implies that the original circuit is modified as follows:

  (a) All voltage sources are short-circuited and current sources open-circuited.
  (b) Each capacitor is replaced by a current source of value $Cm_{i-1}(V_c)$, where $m_{i-1}(V_c)$ is the $(i-1)^{\text{th}}$ moment of the voltage V_c across the capacitor.
  (c) Each self or mutual inductance L_{ij} is replaced by a voltage source of value $L_{ij}m_{i-1}(I_j)$ on line i, where $m_{i-1}(I_j)$ is the $(i-1)^{\text{th}}$ moment of the current through inductor j.

Example – An RC Line: We will demonstrate the above ideas on the RC line shown in Figure 3.3. To simplify our calculations, we will assume that all resistors are of value 1Ω, and all capacitors have the (unrealistic) value of 1F. The

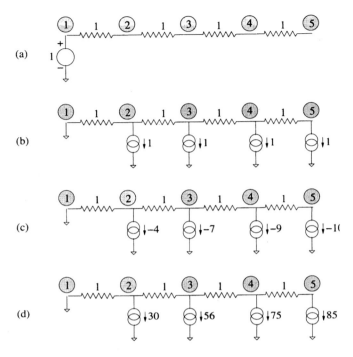

Figure 3.11. Moment calculation for the RC line in Figure 3.3 involves the solution of these circuits. The computation of the zeroth through the third moment requires the solution of the circuits (a) through (d), respectively.

set of unknowns here is characterized by the vector $\mathbf{X} = [v_0 \; v_1 \; v_2 \; v_3 \; v_4 \; v_5 \; i_v]^T$, where v_i is the voltage at node i, and i_v is the current through the voltage source.

Finding $\mathbf{m_0}$ All capacitors are open-circuited, and the voltage source is replaced by a delta function in the time domain, which corresponds to a unit source in the s domain. The resulting circuit is shown in Figure 3.11(a), and can be analyzed by inspection. It is easily verified that no current can flow in the circuit, which implies that the voltage at each node is 1. Therefore,

$$\mathbf{m_0} = \begin{bmatrix} 1 & 1 & 1 & 1 & 1 & 0 \end{bmatrix}^T \tag{3.38}$$

Finding $\mathbf{m_1}$ The capacitor from node i to ground, for $i = 2, 3, 4, 5$, is now replaced by a current source of value $C_i \mathbf{m_0}(V_i)$, and the input is grounded. In this case, it turns out that all of these current sources have unit value, as illustrated in Figure 3.11(b). Again, this circuit can be analyzed by inspection. The current in each branch can be calculated as the sum of all downstream currents, and the voltage at each node may then be calculated to give

$$\mathbf{m_1} = \begin{bmatrix} 0 & -4 & -7 & -9 & -10 & -4 \end{bmatrix}^T \tag{3.39}$$

Finding $\mathbf{m_2}$ *and* $\mathbf{m_3}$ The same technique may be used, in each case replacing a capacitor by a current source of value $C_i\mathbf{m}_{i-1}(V_i)$ and grounding the input excitation. The reader is invited to verify that this leads to:

$$\mathbf{m_2} = \begin{bmatrix} 0 & 30 & 56 & 75 & 85 & 30 \end{bmatrix}^T \tag{3.40}$$

$$\mathbf{m_3} = \begin{bmatrix} 0 & -246 & -462 & -622 & -707 & -246 \end{bmatrix}^T \tag{3.41}$$

Moments of order higher than 3 can be computed in a similar manner.   □

The above example illustrates a few points that can be used to make the following generalizations.

1. *Any* RLC tree or mesh structure driven by a voltage source, by definition, has no resistors or inductors to ground. Therefore, the value of m_0 for all node voltages is 1, since the circuit corresponding to $\mathbf{m_0}$ has no path that allows current to flow. This is merely a corroboration of a result that was shown earlier in this chapter using the final value theorem.

2. During the calculation of the value of the first voltage moment, m_1, for any node in an uncoupled RC tree, the current in each branch is easily verified to be the sum of all downstream capacitance values. This, coupled with the earlier proof that the Elmore delay at a node is the negative of the first voltage moment, leads to an alternative derivation of the Elmore delay formula in Equation (3.17).

3. For an (uncoupled) RLC tree, the current in each branch is given by the sum of all downstream currents. These may therefore be calculated by a simple tree traversal, starting from the sinks and moving towards the source. A second traversal, starting from the source and moving towards the sinks, may then be used to calculate the node voltages, which correspond to the voltage moments. The cost of moment computation for an RC tree is therefore linear in the number of nodes in the tree.

4. For coupled RLC trees, the moment computation rules may be followed to also permit the moments to be computed using tree traversals, in linear time.

Example – A distributed line: The distributed RC line, shown in Figure 3.12, has length L and is characterized by its per unit resistance r and per unit capacitance c. An infinitesimal element of size dx has a resistance of $r \cdot dx$ and a capacitance of $c \cdot dx$.

Moment computation for this line can proceed as usual: for the zeroth moment, the capacitances are all open-circuited, and a unit voltage source is applied. Clearly, the voltage everywhere on the line will be 1, and this yields $m_0 = 1$ everywhere on the line.

To compute m_1, all of the infinitesimal capacitors are replaced by current sources of value $c \cdot dx$ from the line to ground, so that the voltage at any point that is a distance x from the source is given by

$$V(x) = \int_{x=0}^{L} -I(x)(r \cdot dx) \tag{3.42}$$

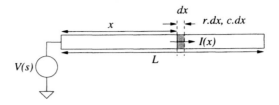

Figure 3.12.   An example of a distributed RC line.

where $I(x)$ is the current passing through the infinitesimal element. This is merely the sum of the downstream currents at point x, and may be computed as $\int_x^L c \cdot dx = c \cdot (L - x)$. Therefore, the value of the first moment, which is identical to $V(x)$, is

$$m_1(x) = rc \left(\frac{x^2}{2} - Lx \right) \tag{3.43}$$

Higher order moments can similarly be calculated to find the moments of the impulse response at a point that is x units from the source as $H(x) = 1 + m_1 s + m_2 s^2 + m_3 s^3 + \cdots$.

Example – A coupled RLC line: A simple coupled RLC line is illustrated in Figure 3.13(a). The coupling elements correspond to a floating capacitance C_c, and mutual inductances $L_{12} = L_{21}$, and line i has a resistive element of R_i, a grounded capacitance of C_i, and a self-inductance of L_{ii}.

The equivalent circuit for the zeroth moment is shown in Figure 3.13(b), and it is easy to see from this that the voltage moments everywhere on line1 are 1, those everywhere in line2 are 0, and the current moments everywhere are 0. This yields the circuit for the first moment calculation in Figure 3.13(c). Observe that since the zeroth current moments are all zero, the voltage source corresponding to each inductor is zero. As a result, inductors do not affect m_1 in any way. The m_1 voltage moments at the three nodes on line1 are, in order from the distance from the source, 0, -2, and -2, and the current moment for line 1 is -2. The corresponding numbers for line2 are voltage moments of 0, 1 and 1, and a current moment of 1.

The circuit for the computation of m_2 is depicted in Figure 3.13(d). Observe that since the first current moments on both lines were nonzero, the inductances will begin to affect the moments from m_2 onwards.

As before, this procedure may be used to find any arbitrary moment m_i, and moment matching may be used to find a Padé approximant.

3.5.3   Transfer function approximation using moment matching

Having calculated the moments of the transfer function, the next step is to find a Padé approximant of order $[L/M]$. For a physical system, we must have $L \leq M$ for the voltage transfer function. The inequality is strict for a system with zero initial conditions, and the case where $L = M$ corresponds to a system with nonzero initial voltages or currents[6].

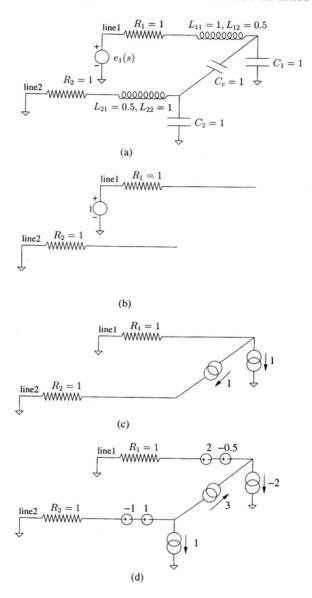

Figure 3.13. Moment calculation for a coupled RLC line. The original circuit is shown in (a), and (b), (c) and (d) depict the equivalent circuits for the computation of m_0, m_1 and m_2, respectively.

The process of *moment matching* was informally introduced earlier, and we now discuss more formally how this may be used to find the Padé approximant of a given order, say, $[(q-1)/q]$. We begin with the moments of a response, such as the voltage moments of a node in the circuit, and attempt to set

$$m_0 + m_1 s + m_2 s^2 + \cdots \equiv \frac{a_0 + a_1 s + \cdots + a_{q-1} s^{q-1}}{1 + b_1 s + \cdots + b_q s^q} \qquad (3.44)$$

There are $2q$ unknowns here, corresponding to the a_i and b_i variables, and therefore, we must generate a system of $2q$ equations to find their values. Cross-multiplying the denominator of the right hand side with the left-hand side, this may be written as

$$(m_0 + m_1 s + m_2 s^2 + \cdots) \cdot (1 + b_1 s + \cdots + b_q s^q) \equiv a_0 + a_1 s + \cdots + a_{q-1} s^{q-1} \quad (3.45)$$

Matching the powers of s on either side, we obtain the following equations from the first q powers of s.

$$
\begin{aligned}
s^0 : & \quad a_0 &=& \quad m_0 \\
s^1 : & \quad a_1 &=& \quad m_1 + m_0 b_1 \\
s^2 : & \quad a_2 &=& \quad m_2 + m_1 b_1 + m_0 b_2
\end{aligned}
\qquad (3.46)
$$

$$
\vdots
$$

$$
s^{q-1} : \quad a_{q-1} \quad = \quad m_{q-1} + m_{q-2} b_1 + \cdots m_1 b_{q-2} + m_0 b_{q-1}
$$

$$(3.47)$$

For the next q powers of s, the left hand side is zero, and we obtain the following equations

$$
\begin{aligned}
s^q : & \quad 0 &=& \quad m_q + m_{q-1} b_1 + \cdots m_1 b_{q-1} + m_0 b_q \\
s^{q+1} : & \quad 0 &=& \quad m_{q+1} + m_q b_1 + \cdots m_2 b_{q-1} + m_1 b_q
\end{aligned}
\qquad (3.48)
$$

$$
\vdots
$$

$$
s^{2q-1} : \quad 0 \quad = \quad m_{2q-1} + m_{2q-2} b_1 + \cdots m_q b_{q-1} + m_{q-1} b_q
$$

The equations represented in (3.48) can be written more compactly as the set of linear equations:

$$
\begin{bmatrix}
m_0 & m_1 & m_2 & \cdots & m_{q-2} & m_{q-1} \\
m_1 & m_2 & m_3 & \cdots & m_{q-1} & m_q \\
m_2 & m_3 & m_4 & \cdots & m_q & m_{q+1} \\
\vdots & \vdots & \vdots & \ddots & \vdots & \vdots \\
m_{q-1} & m_q & m_{q+1} & \cdots & m_{2q-3} & m_{2q-2}
\end{bmatrix}
\begin{bmatrix}
b_q \\
b_{q-1} \\
\vdots \\
b_1
\end{bmatrix}
= -
\begin{bmatrix}
m_q \\
m_{q+1} \\
m_{q+2} \\
\vdots \\
m_{2q-1}
\end{bmatrix}
\quad (3.49)
$$

This is a system of q linear equations in q variables that can easily be solved to find the values of the b_i variables. Since the left hand side matrix, which is in the Hankel form, is dense, the computational complexity of this step is $O(q^3)$; however, since q is typically small (typically, $q \leq 5$ for RC circuits), this is not a significant expense. Once the values of the b_i's are known, a simple substitution on the right hand sides of the equations in (3.46) yields the values of the a_i variables.

Once the Padé approximant to the transfer function has been computed, the final step is to compute a response to an applied stimulus. For a unit-step stimulus and an approximation of order $[(q-1)/q]$, this involves finding the inverse Laplace transform of

$$\frac{a_0 + a_1 s + \cdots + a_{q-1}s^{q-1}}{s(1 + b_1 s + \cdots + b_q s^q)} \qquad (3.50)$$

A first step is to find the roots of the denominator, which reduces to finding the roots of the polynomial $1 + b_1 s + \cdots + b_q s^q$. For a polynomial of order $q \le 4$, closed-form solutions for the roots exist; for higher order polynomials, numerical techniques are essential. Specifically, the closed-form formula for the roots of a quadratic equation is widely known and is not reproduced in this book. The formulæ for the roots of cubic and quartic equations are provided in Appendix A. For this reason, and because higher-order AWE approximations are often unstable, it is common to select $q \le 4$. Next, given these poles, the approximant is represented as a sum of partial fractions, and the corresponding residues are calculated. The final step of inverse Laplace transforming this sum yields the response as a sum of exponentials.

A typical approach to finding an reduced order system is to start with a desired order and to compute the corresponding approximant. The stability of this approximant is easily checked using, for example, the Routh-Hurwitz criterion [Kuo93]. If the solution is unstable, the order is successively reduced until a stable solution is found. The first order approximant is guaranteed to be stable for an RLC system, and therefore this procedure will definitely result in a solution; whether the level of accuracy of the solution is satisfactory or not is another matter altogether.

Example: For the RC line shown in Figure 3.3, the moments of the voltage at node 5 were calculated earlier as $1 - 10s + 85s^2 - 707s^3 + \cdots$. Matching these to a $[1/2]$ Padé approximant, we obtain

$$(1 - 10s + 85s^2 - 707s^3 + \cdots)(1 + b_1 s + b_2 s^2) \equiv a_0 + a_1 s \qquad (3.51)$$

which leads to the equations

$$
\begin{aligned}
a_0 &= 1 \\
a_1 &= b_1 - 10 \\
0 &= b_2 - 10b_1 + 85 \\
0 &= -10b_2 + 85b_1 - 707
\end{aligned}
$$

Solving these, we obtain the $[1/2]$ Padé approximant to the transfer function as

$$\frac{1 - \frac{7}{15}s}{1 + \frac{143}{15}s + \frac{31}{3}s^2}$$

A comparison of this with the exact response to a step excitation was shown earlier, in Figure 3.4(b). $\qquad \square$

3.5.4   Moment scaling and moment shifting

Since typical time constants in integrated circuits are of the order of picoseconds to nanoseconds, consecutive moments can differ by many orders of magnitude. Typically, the ratio of consecutive moments is of the order of the dominant time constant of the system. This may lead to numerical errors due to the limited precision afforded by computing machines. This is effectively addressed by the use of *moment scaling*. If all capacitors and inductors are scaled by a multiplicative factor, k, this is equivalent to multiplying the frequency variable, and hence the time variable, by a factor of $1/k$. For instance, if k is set to 10^9, then the only difference after scaling is that the units of time are altered from seconds to nanoseconds; the numerical errors in the computation are, however, greatly reduced.

The idea of *moment shifting* is to compute moments about a point other than $s = 0$, say about $s = \alpha$, since this may be able to better capture the effects of nondominant poles. It is easy to show that such a shift is equivalent to adding a conductance of αC in parallel with a capacitor of value C, and a resistance of αL in series with an inductor of value L; for a detailed discussion, the reader is referred to [CPO02].

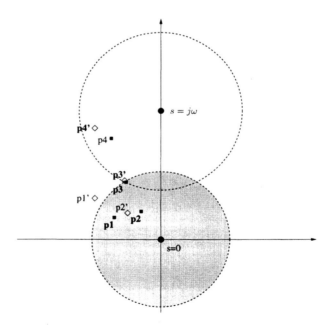

Figure 3.14.   An illustration of complex frequency hopping. The boldfaced poles correspond to those that lie inside the radius of convergence and are deemed correct (adapted from [CHNR93]).

3.5.5   Limitations of AWE

The AWE technique is based on the MacLaurin series expansion, and is therefore limited by the region of convergence of this approximation. A significant limitation of this method is that the Padé approximant for a stable system may very well be unstable, as the approximant creates a spurious pole in the right half plane. While a first-order approximation for an RLC circuit is provably stable, no such guarantee exists for higher order expansions. However, in practice, it often does provide effective and stable solutions for many interconnect systems in digital circuits.

Several techniques have been employed to enhance the stability of AWE, one among which is the notion of complex frequency hopping [CHNR93]. In this procedure, illustrated in Figure 3.14, the Padé approximation about $s = 0$ is first computed, and the corresponding poles p1, p2, p3 and p4 are identified. These are accurate if they fall within the radius of convergence of the expansion; however, it is not possible to say what this radius is. Therefore, a second expansion is carried out about $s = j\omega$, for some wisely chosen value of ω. This expansion results in another set of poles, p1', p2', p3' and p4': some of these may match the original set, while others will not. If any pole is found to match, it must lie within the radius of convergence, and consequently, *all poles* that are closer to the expansion point are deemed accurate. In this case, p3 and p3' coincide. Therefore, the radius of convergence is at least as large as the dotted circles shown in the figure, and therefore, the set of accurate poles that underlined in the figure, include p1, p2 and p3 for the lower circle, and p3' (which is identical to p3) and p4' for the upper circle. In this specific case, all poles were found to be accurate after one hop, but in general, this may not be so. In such a case, one could continue hopping along the complex $(j\omega)$ axis for several expansion points, until all poles are deemed accurate.

While it is possible to perform hopping over the entire complex plane, the right half plane is of no interest in physical systems since it contains no poles, and points in the left half plane that are far from the imaginary axis typically do not constitute dominant poles, and can be excluded. Therefore, in practice, the hopping procedure is only carried out on the imaginary axis. Further computational simplifications are possible. Since complex poles appear in conjugate pairs, it is only necessary to perform complex frequency hopping on the upper (or lower) half of the imaginary axis.

An additional limitation of AWE is related to the inability of the reduced order models to guarantee passivity. A more detailed discussion of this issues is deferred to Section 3.6.3.

3.6   KRYLOV SUBSPACE-BASED METHODS

3.6.1   Numerical stability and Krylov subspaces

The AWE method can be very useful in building reduced order systems, but there are several issues associated with the quality and numerical properties of the solution. While the latter may be partially overcome using moment scaling in some circumstances, such an approach provides limited relief. The use of Krylov subspace-based methods provides an effective way to overcome a number of problems associated with numerical instability.

We begin by defining the span of a set of vectors, which in turn is used to define a Krylov subspace

Definition 3.6.1 *The span of a set of n-dimensional vectors* $\mathbf{v}_1, \mathbf{v}_2, \cdots, \mathbf{v}_k$ *is the set of points that can be expressed as a linear combination of these vectors. In other words,* $\mathbf{v} \in span\{\mathbf{v}_1, \mathbf{v}_2, \cdots, \mathbf{v}_k\}$ *implies that* $\mathbf{v} = \sum_{i=1}^{k} \alpha_i \mathbf{v}_i$ *for some set of real values* α_i.

Definition 3.6.2 *A Krylov subspace of order q associated with an* $n \times n$ *matrix A and a vector* $\mathbf{r}$, *denoted by* $\mathcal{K}_q(A, \mathbf{r})$ *is given by the span of the vectors*

$$\{\mathbf{r}, A\mathbf{r}, A^2\mathbf{r}, \cdots, A^{q-1}\mathbf{r}\} \tag{3.52}$$

Interestingly, it can be proven that the sequence $A^{q-1}\mathbf{r}$ converges to the eigenvector corresponding to the largest eigenvalue of A as q becomes large, regardless of the chosen (nonzero) value of $\mathbf{r}$.

If the vector $\mathbf{r}$ is replaced by a $n \times p$ matrix R (where typically $p \ll n$), then the subspace $\mathcal{K}_q(A, R)$ is similarly defined as the span of the columns of the matrices $\{R, AR, A^2R, \cdots, A^{q-1}R\}$, and is referred to as a **block Krylov subspace**.

The relationship between the moment generation process and Krylov subspaces can be seen from Equations (3.36) and (3.37). From the latter, it is easy to see that the sequence of moments

$$\mathbf{m}_0, \mathbf{m}_1, \mathbf{m}_2, \cdots, \mathbf{m}_i, \cdots$$

can be rewritten as

$$\mathbf{m}_0, (G^{-1}C)\mathbf{m}_0, (G^{-1}C)^2\mathbf{m}_0, \cdots, (G^{-1}C)^k\mathbf{m}_0, \cdots$$

These vectors together match the above definition of a Krylov subspace, where $A = -G^{-1}C$ and $\mathbf{r} = \mathbf{m}_0$, and we will henceforth use A and $\mathbf{r}$ to denote these terms.

Instead of working with $A^i\mathbf{r}$ directly, which is liable to introduce numerical errors, we will work with an *orthonormal basis* matrix within the subspace: for a Krylov subspace $\mathcal{K}_q$, an orthonormal basis $\{\mathbf{v}_1, \mathbf{v}_2, \cdots, \mathbf{v}_k\}$ consists of a set of vectors $\mathbf{v}_i, 1 \leq i \leq k$ such that $span\{\mathbf{v}_1, \mathbf{v}_2, \cdots, \mathbf{v}_k\} = \mathcal{K}_q(A, \mathbf{r})$ and the vectors $\mathbf{v}_i$ are all orthogonal, i.e., $\mathbf{v}_i^T \mathbf{v}_j = 0 \ \forall \ i \neq j$ and $\mathbf{v}_i^T \mathbf{v}_i = 1 \ \forall \ i$.

Techniques such as Lanczos-based [FF95, FF96, FF98, Fre99] and Arnoldi-based methods [SKEW96, EL97] overcome the numerical limitations associated with practical implementations of AWE at the expense of slightly more complex calculations. Other methods such as truncated balanced realizations have also been studied recently [PDS03] to overcome some of the limitations in accuracy of other methods, and to more easily widen their applicability to a larger class of circuits.

3.6.2   The block Arnoldi method

A block Arnoldi method for solving circuit equations is proposed in [SKEW96]. We present a description of the procedure based on the explanation in [OCP98]. Consider a circuit whose inputs and outputs are all considered to be the p ports of the system (we will show a specific example in Section 3.6.3). Its equations are given by

$$(G + sC)\mathbf{X} = B\mathbf{u} \tag{3.53}$$
$$\mathbf{y} = L^T\mathbf{X}$$

where G and C are $n \times n$ matrices, B and L are $n \times m$ matrices, and $\mathbf{X}$, $\mathbf{u}$ and $\mathbf{y}$ are column vectors of dimension n, m and p, respectively. This is essentially similar to the descriptions used earlier: the first of these equations is almost identical to Equation (3.33). The second, the output equation, represents the values at the m ports (including the outputs of interest) contained in $\mathbf{y} = [y_1 \cdots y_p]^T$, which are represented as linear functions of the $\mathbf{X}$ variables: for instance, if the output of interest is the kth element of $\mathbf{X}$, then $y_k = 1$ and $y_i = 0$ for all other i.

From Equation (3.53), simple matrix algebraic techniques can be used to obtain

$$\mathbf{y} = Y\mathbf{u} \tag{3.54}$$
$$\text{where } Y = L^T(G + sC)^{-1}B \tag{3.55}$$
$$= L^T(I - sA)^{-1}R \tag{3.56}$$

where $A = -G^{-1}C$ as before, and $R = G^{-1}B$. This may be expanded about $s = 0$ to obtain

$$Y = L^T R - sL^T AR + s^2 L^T A^2 R + \cdots + s^i L^T A^i R + \cdots \tag{3.57}$$

or in other words, the $p \times p$ matrix representing the ith moment $M_i = L^T A^i R$. Note that the entry in the position (a, b) of matrix M_i corresponds to the ith moment of transfer function that relates the ath entry of $\mathbf{y}$ with the bth entry of $\mathbf{u}$.

Instead of using moments, for numerical stability, an Arnoldi-based method generates an orthonormal basis $\mathbf{\Psi}$ for the block Krylov subspace $\mathcal{K}_q(A, R)$. The following properties of $\mathbf{\Psi}$ follow as a natural consequence:

$$span(\mathbf{\Psi}) = \mathcal{K}_q(A, R) \tag{3.58}$$
$$\mathbf{\Psi}^T\mathbf{\Psi} = I \tag{3.59}$$

where I is the identity matrix.

For a qth order reduction, an Arnoldi-based method uses the $n \times q$ matrix $\mathbf{\Psi}$ matrix to apply the variable transformation:

$$\mathbf{X} = \mathbf{\Psi}\tilde{\mathbf{x}} \tag{3.60}$$

This transformation effectively maps a point $\tilde{\mathbf{x}}$ in a q-dimensional space to a point $\mathbf{X}$ in the original n-dimensional space. Substituting this in (3.53), premultiplying the first equation by $\mathbf{\Psi}^T G^{-1}$, and using Equation (3.59), we obtain

$$\tilde{\mathbf{x}} + s\mathbf{\Psi}^T A\mathbf{\Psi}\tilde{\mathbf{x}} \ = \ \mathbf{\Psi}^T R\mathbf{u} \tag{3.61}$$
$$y \ = \ L^T \mathbf{\Psi}\tilde{\mathbf{x}}$$

Note that this is a reduced system since it is represented by q state variables instead of the original n variables. This may be rewritten as

$$\tilde{\mathbf{x}} \ = \ sH_q\tilde{\mathbf{x}} + \mathbf{\Psi}^T R\mathbf{u} \tag{3.62}$$
$$y \ = \ L^T \mathbf{\Psi}\tilde{\mathbf{x}}$$

where we set $H_q = \mathbf{\Psi}^T A\mathbf{\Psi}$, a $q \times q$ matrix that corresponds to the reduced order system. In the Laplace domain, this implies that the transfer function for this reduced-order system is given by

$$\tilde{Y} \ = \ L^T \mathbf{\Psi}(I - sH_q)^{-1}\mathbf{\Psi}^T R \tag{3.63}$$

Using the eigendecomposition $H_q = S\Lambda_q S^{-1}$, we obtain

$$\tilde{Y} \ = \ L^T \mathbf{\Psi}S(I - s\Lambda_q)^{-1}S^{-1}\mathbf{\Psi}^T R \tag{3.64}$$

The matrix $(I - s\Lambda_q)$ is diagonal and is hence easily invertible for small values of the approximation order, q. As a result, this method is computationally efficient. Since it does not directly work with circuit moments, it succeeds in avoiding the numerical problems faced by AWE.

3.6.3 Passivity and PRIMA

Numerical problems are not the only issues that haunt AWE-like methods. A major problem associated with solutions provided by AWE and simple Arnoldi- or Lanczos-based methods is that they cannot guarantee that the resulting network is *passive*. A passive network is one that always dissipates more energy than it generates; pure RLC networks always possess this property. Stability (which is equivalent to the statement that all poles of the transfer function must lie in the left half plane) is a necessary *but not sufficient* condition for passivity, so that of the two, passivity is the stronger and more restrictive condition.

In particular, it is possible to show existence cases of a nonpassive but stable system that, when connected to a passive and stable system, results in an unstable system [KY98]. On the other hand, if a system is guaranteed to be passive, any interconnection with any other passive system will also be passive (and hence stable). This idea was described in the context of model order reduction in [KY98], and the idea of a split congruence transformation was proposed. This was later incorporated into other model order reduction algorithms, among which PRIMA is widely used.

The Passive Reduced-order Interconnect Macromodeling Algorithm (PRIMA) [OCP98] algorithm, based on the idea of split congruence transformations to ensure passivity and block Krylov subspace methods to assure numerical stability, overcomes many of the shortcomings of AWE and provides a guaranteed stable and passive reduced order model for the linear system. This comes at the expense of (typically) requiring a model of higher order to achieve the same accuracy as AWE (assuming, of course, that the AWE model is stable). From the programmer's point of view, a major advantage of the PRIMA approach is that it is very similar to AWE in terms of implementation, and requires a few additional intermediate steps.

An Arnoldi-based implementation of PRIMA, which guaranteed the passivity of the reduced solution, can be described as follows. In terms of equation formulation, it has been shown in [OCP98] that if the following two criteria are satisfied, the representation will result in a passive reduced order model:

1. The MNA equations in (3.53) are written in such a way that

$$
C = \left[\begin{array}{ccc} Q & \vdots & 0 \\ \cdots\cdots\cdots & & \\ 0 & \vdots & H \end{array} \right] \quad G = \left[\begin{array}{ccc} N & \vdots & E \\ \cdots\cdots\cdots\cdots & & \\ -E^T & \vdots & 0 \end{array} \right] \tag{3.65}
$$

where resistor stamps are included in N, capacitor stamps in Q, and Type 2 element stamps are in the border elements. For example, voltage source stamps manifest themselves in E, and inductors, which are treated as Type 2 elements, contribute their stamps to H and E.

2. To guarantee passivity, it is necessary to set $B = L$, which requires some special handling during equation formulation.

Example: For the circuit shown in Figure 3.3, if the parameter of interest is the voltage at node 5, then the circuit equations may be written in the following manner to satisfy the above requirements:

$$
\left(\begin{bmatrix} 1 & -1 & 0 & 0 & 0 & \vdots & 1 \\ -1 & 2 & -1 & 0 & 0 & \vdots & 0 \\ 0 & -1 & 2 & -1 & 0 & \vdots & 0 \\ 0 & 0 & -1 & 2 & -1 & \vdots & 0 \\ 0 & 0 & 0 & -1 & 1 & \vdots & 0 \\ \cdots\cdots\cdots\cdots\cdots\cdots\cdots \\ -1 & 0 & 0 & 0 & 0 & \vdots & 0 \end{bmatrix} + s \begin{bmatrix} 0 & 0 & 0 & 0 & 0 & \vdots & 0 \\ 0 & 1 & 0 & 0 & 0 & \vdots & 0 \\ 0 & 0 & 1 & 0 & 0 & \vdots & 0 \\ 0 & 0 & 0 & 1 & 0 & \vdots & 0 \\ 0 & 0 & 0 & 0 & 1 & \vdots & 0 \\ \cdots\cdots\cdots\cdots\cdots\cdots\cdots \\ 0 & 0 & 0 & 0 & 0 & \vdots & 0 \end{bmatrix} \right) \begin{bmatrix} V_1 \\ V_2 \\ V_3 \\ V_4 \\ V_5 \\ \cdots \\ I_V \end{bmatrix}
$$

$$
= \begin{bmatrix} 0 & 0 \\ 0 & 0 \\ 0 & 0 \\ 0 & 0 \\ 0 & 1 \\ -1 & 0 \end{bmatrix} \begin{bmatrix} V_s \\ I_{dummy} \end{bmatrix} \quad (3.66)
$$

$$
\begin{bmatrix} y_{dummy} \\ y_{out} \end{bmatrix} = \begin{bmatrix} 0 & 0 & 0 & 0 & 0 & -1 \\ 0 & 0 & 0 & 0 & 1 & 0 \end{bmatrix} \begin{bmatrix} V_1 \\ V_2 \\ V_3 \\ V_4 \\ V_5 \\ I_V \end{bmatrix} \quad (3.67)
$$

Note the negation of both sides of the device equation for the voltage source to maintain the negative transpose relationship relationship between the off-diagonal blocks of the MNA matrix, and the addition of an extra dummy current source, I_{dummy}, whose primary function is to ensure the relationship $B = L$ (this also results in $m = p$). Similarly, a dummy output y_{dummy} is forced upon the formulation to maintain this relationship. Once the reduced order model is generated, we set $I_{dummy} = 0$.   □

As in the block Arnoldi case, PRIMA generates an orthonormal basis $\boldsymbol{\Psi}$ for the block Krylov subspace $\mathcal{K}_q(A, R)$, with the properties described in Equations (3.58) and (3.59), and uses the variable transformation $\mathbf{X} = \boldsymbol{\Psi}\tilde{\mathbf{x}}$.

This is now substituted in (3.53), but the first equation is now premultiplied instead by $\boldsymbol{\Psi}^T$ (instead of $\boldsymbol{\Psi}^T G^{-1}$, as in Section 3.6.2) to yield

$$
\begin{aligned}
\boldsymbol{\Psi}^T G \boldsymbol{\Psi}\tilde{\mathbf{x}} + \boldsymbol{\Psi}^T sC\boldsymbol{\Psi}\tilde{\mathbf{x}} &= \boldsymbol{\Psi}^T B\mathbf{u} \qquad (3.68) \\
\mathbf{y} &= L^T \boldsymbol{\Psi}\tilde{\mathbf{x}}
\end{aligned}
$$

For this reduced system of order q, we substitute $\tilde{G} = \boldsymbol{\Psi}^T G \boldsymbol{\Psi}$, $\tilde{C} = \boldsymbol{\Psi}^T C \boldsymbol{\Psi}$, $\tilde{L} = \boldsymbol{\Psi}^T L$, and $\tilde{B} = \boldsymbol{\Psi}^T B$ to obtain the following representation of the reduced

1. **Algorithm PRIMA**

2. Find $R = G^{-1}B$

   /* Gram-Schmidt orthonormalization of the columns of R */

3. $\Psi_0 = GS(R, p)$

4. Set $n = \lceil \frac{q}{N} \rceil$

5. /* Gram-Schmidt orthonormalization of the columns of Ψ_k^k */
   for $k = 1$ to n do

        Solve $G\Psi_k^0 = C\Psi_{k-1}$ for Ψ_k^0

        $\Psi_k = GS(\Psi_k^0, kp)$

6. $\Psi = $ first q columns of $[\Psi_0 \Psi_1 \cdots \Psi_n]$

7. Compute $\tilde{C} = \Psi^T C \Psi$ $\tilde{G} = \Psi^T G \Psi$, $\tilde{H}_q = \tilde{G}^{-1}\tilde{C}$

8. Eigendecompose $\tilde{H}_q = S\Lambda_q S^{-1}$ where $\Lambda_q = diag(\lambda_1, \lambda_2, \cdots, \lambda_q)$

9. To find the poles and residues of $y_{i,j}(s)$

        /* $B = [\mathbf{B}_1 \ \mathbf{B}_2 \ \cdots \ \mathbf{B}_n]$ */

        Find $\mathbf{w} = \tilde{G}^{-1}\Psi^T \mathbf{B}_j$

        Find $\mu = S^T\Psi^T \mathbf{1}_i$, $\nu = S^{-1}\mathbf{w}$

        /* $\mu = [\mu_1 \ \mu_2 \ \cdots \ \mu_n]^T$, $\nu = [\nu_1 \ \nu_2 \ \cdots \ \nu_n]^T$ */

        $y_{i,j}(s) = \sum_{k=1}^q \frac{\mu_k \nu_k}{1 - s\lambda_k}$

10. **end PRIMA**

11. **Algorithm GS(X,n)**

    /* $X = [\mathbf{X}_1 \ \mathbf{X}_2 \ \cdots \ \mathbf{X}_n]$ */

12. $\mathbf{Y}_1 = \frac{\mathbf{X}_1}{\|\mathbf{X}_1\|}$

13. for $i = 1$ to n do

         for $j = 1$ to $i - 1$ do

              $\mathbf{Y}_i = \mathbf{X}_i - (\mathbf{X}_i^T \mathbf{Y}_j)\mathbf{Y}_j$

         $\mathbf{Y}_i = \frac{\mathbf{Y}_i}{\|\mathbf{Y}_i\|}$

14. **end GS**

Figure 3.15.   Pseudocode for the PRIMA algorithm.

system:

$$(\tilde{G} + s\tilde{C})\tilde{\mathbf{x}} = \tilde{B}\mathbf{u} \tag{3.69}$$
$$\mathbf{y} = \tilde{L}^T\tilde{\mathbf{x}}$$

We define the $q \times q$ matrix, $\tilde{H}_q = \tilde{G}^{-1}\tilde{C}$, and the solution to the reduced system is given, in a similar manner to Equation (3.56), by

$$\tilde{\mathbf{y}} = \tilde{L}^T(I - s\tilde{H}_q)^{-1}\tilde{R} \tag{3.70}$$

where $\tilde{R} = \tilde{G}^{-1}\tilde{B}$. Since this is a qth order system and q is manageably small, it can easily be solved exactly. Specifically, we eigendecompose the matrix $\tilde{H}_q = S\tilde{\Lambda}_q S^{-1}$, where $\tilde{\Lambda}_q = diag(\lambda_1, \lambda_2, \cdots, \lambda_q)$, to obtain

$$\tilde{\mathbf{y}} = \tilde{L}^T S(I - s\Lambda_q)^{-1}S^{-1}\tilde{R} \tag{3.71}$$

The matrix $(I - s\tilde{\Lambda}_q)$ is diagonal and is invertible in $O(q)$ time, and therefore the right hand side can be evaluated.

The pseudocode for PRIMA is provided in Figure 3.15. Additional practical tips for increasing the robustness of the method have been provided in [OCP99]. **Example:** For the example in Figure 3.3, the first step of the algorithm implies solving $GR = B$, which is equivalent to the original system twice at dc, each time with a unit source excitation on one of the two sources. Instead of finding $G^{-1}B$ using matrix multiplication, this may be solved by short-circuiting all inductors (this is irrelevant for this circuit, since it has none), open-circuiting all capacitors, and applying these excitations. The resulting circuit can be solved using path traversals to obtain

$$R = \begin{bmatrix} 1 & 1 & 1 & 1 & 1 & 0 \\ 0 & 1 & 2 & 3 & 4 & 1 \end{bmatrix}^T$$

When this is orthonormalized, we obtain

$$\Psi_0 = \begin{bmatrix} -0.4472 & -0.4472 & -0.4472 & -0.4472 & -0.4472 & 0 \\ -0.6030 & -0.3015 & 0 & 0.3015 & 0.6030 & 0.3015 \end{bmatrix}^T$$

Next, we apply the loop in line 4 of the pseudocode and at the end of step 5, we calculate Ψ as

$$\Psi = \begin{bmatrix} -0.4472 & -0.6030 & -0.3088 & -0.0210 \\ -0.4472 & -0.3015 & 0.2702 & -0.4631 \\ -0.4472 & 0 & 0.4247 & 0.1894 \\ -0.4472 & 0.3015 & 0.1544 & 0.6525 \\ -0.4472 & 0.6030 & -0.5405 & -0.3578 \\ 0 & 0.3015 & 0.5791 & -0.4420 \end{bmatrix}$$

This leads to

$$\tilde{G} = \begin{bmatrix} 0 & -0.1348 & -0.2590 & 0.1977 \\ 0.1348 & 0.3636 & -0.3259 & 0.1714 \\ 0.2590 & 0.1862 & 0.9151 & 0.5704 \\ -0.1977 & -0.3744 & 0.2730 & 1.8564 \end{bmatrix}$$

$$
\tilde{C} = \begin{bmatrix} 0.8 & -0.2697 & -0.1381 & -0.0094 \\ -0.2697 & 0.5455 & -0.3608 & 0.1206 \\ -0.1381 & -0.3608 & 0.5693 & 0.2495 \\ -0.0094 & 0.1206 & 0.2495 & 0.8042 \end{bmatrix} , \tilde{L} = \tilde{B} = \begin{bmatrix} 0 & -0.4472 \\ -0.3015 & 0.6030 \\ -0.5791 & -0.5405 \\ 0.4420 & 0.3578 \end{bmatrix}
$$

Taking the algorithm further, we find the fourth order approximation to the voltage transfer function in the s domain at node 5 as

$$
\frac{3.5797}{1 - 8.2808s} + \frac{0.6737}{1 - 0.9616s} + \frac{0.0222}{1 - 0.0155s} + \frac{-0.2095}{1 - 0.4263s}
$$

It is easy to verify that the sixth order approximation returns the original system.   □

3.7   FAST DELAY METRICS

3.7.1   The Elmore delay as an upper bound

The step response for an RC tree can be proven to increase monotonically from 0 to 1. In this sense, it behaves in a similar manner as a cumulative distribution function (cdf) of a probability distribution of some random variable, and several of the results from probability theory may be brought to bear in this regard. For example,

- The impulse response, which is the derivative of the step response, can be thought of as being equivalent to a probability density function (pdf) of the same random variable.

- The Elmore delay, which is the first time moment of the impulse response, is equivalent to the mean value of the random variable.

- Higher order moments, such as the second and third moments, can be countenanced as being equivalent to their corresponding counterparts in probability theory.

A specific concept that is widely used in probability theory is that of a *central moment*. The kth central moment of a probability distribution on a random variable x described by a density function $f(x)$ is given by

$$
\int_{-\infty}^{\infty} (x - \overline{x})^k f(x) dx \tag{3.72}
$$

where $\overline{x}$ is the mean of $f(x)$. Since the impulse response $h(t)$ of a system can be interpreted as a probability density function, we may write its kth central moment as

$$
\int_{0}^{\infty} (t - \overline{t})^k h(t) dx \tag{3.73}
$$

where $\overline{t}$ is the first moment, or mean, of $h(t)$. The relationship between the moments m_i, as defined in Equation (3.24), and the first few central moments,

μ_i, is given by

$$\mu_0 = 1 \tag{3.74}$$
$$\mu_1 = 0$$
$$\mu_2 = 2m_2 - m_1^2$$
$$\mu_3 = -6m_3 + 6m_1 m_2 - 2m_1^3$$

While the values of μ_0 and μ_1 are obvious from earlier discussions, the others have specific interpretations. The second central moment, μ_2, corresponds to the *variance* of the distribution, which is higher if $h(t)$ has a wider "spread" from its mean value. The third central moment, μ_3, has an interpretation as the *skewness* of the distribution: its sign is positive [negative] if the mode, or the maximum of $h(t)$ is to the left [right] of the mean, and its magnitude measures the distance from the mean. The fourth central moment is the *kurtosis* of the distribution, which reflects the total area that corresponds to the tails of $h(t)$.

The above parameters provide measures for the mean and the mode of the distribution; however, the 50% delay that we are interested in is the *median* of the distribution, and generally speaking, no simple closed-form formulas for the median are available. However, it was proved in [GTP97] that for the impulse response $h(t)$ at any node of an RC tree, the Elmore delay, which corresponds to the mean, is an upper bound on the median, or the 50% delay.

3.7.2   Delay metrics based on probabilistic interpretations

Several techniques have used the probabilistic interpretation of moments to arrive at a delay metric, and these are surveyed here. All of the methods listed here are stable in that they result in reasonable (i.e., physically realizable) values for the parameters that define the delay distributions, and therefore, physically reasonable delay values.

The PRIMO metric. The PRIMO method [KP98] attacked the problem by fitting the impulse response to the gamma distribution, given by

$$g_{\lambda,n}(t) = \frac{\lambda^n t^{n-1} e^{-\lambda t}}{\Gamma(n)} \tag{3.75}$$
$$\text{where } \Gamma(x) = \int_0^\infty y^{x-1} e^{-y} dy$$

This distribution is completely described by two parameters, λ and n, and therefore a gamma function approximation to $h(t)$ can be found by fitting the moments of the gamma function to the moments of $h(t)$. The first few central moments of $g_{\lambda,n}(t)$ are given by

$$\mu_1 = 0$$
$$\mu_2 = \frac{n}{\lambda^2}$$
$$\mu_3 = \frac{2n}{\lambda^3}$$

This leads to the relations

$$\lambda = \frac{2\mu_2}{\mu_3} \tag{3.76}$$

$$n = \frac{4\mu_2^3}{\mu_3^2} \tag{3.77}$$

While matching the second and third central moments would certainly yield values of n and λ, in practice, it was found that a time-shifted gamma function provided more accurate results. Essentially, this involves matching the first moment to a time shift, Δ; since the first moment of $g_{\lambda,n}(t)$ is $\frac{n}{\lambda}$, this results in setting

$$\Delta = m_1 - \frac{n}{\lambda} \tag{3.78}$$

While closed forms for the median do not exist, it is relatively simple to store them in two-dimensional look-up tables that are functions of n and λ. If the input is not a step, but can be modeled as a saturated ramp (which can be written as the difference between two time-shifted ramp functions), the response can be captured by two time-shifted gamma functions, separated by the rise time of the saturated ramp. For details, the reader is referred to [KP98].

The h-gamma metric.   This technique was found to have limited accuracy in cases such as those when interconnect resistance plays a large part, and a modification, called the h-gamma method, was proposed in [LAP98]. The method is predicated on the observation that the Laplace transform of the step response can be divided into two components: a forced response and a homogeneous response, as shown below, and illustrated in Figure 3.16:

$$
\begin{aligned}
E(s) &= \frac{1}{s}H(s) \\
&= \frac{m_0}{s} + m_1 + m_2 s + m_3 s^2 + \cdots \\
e(t) &= m_0 u(t) + m_1 y_h(t)
\end{aligned}
\tag{3.79}
$$

where $m_0 u(t)$ is the forced response ($u(t)$ is the unit step function) and $m_1 y_h(t)$ is the homogeneous response, $e(t)$ and $E(s)$ are the step response in the time and frequency domains, respectively, and $H(s)$ is the transfer function in the Laplace domain.

For an RC mesh, the homogeneous response increases monotonically from 0 to 1 under a step excitation, and the h-gamma method treats $y_h(t)$ as a probability density function that is approximated by the gamma distribution. Specifically, in the frequency domain, the homogeneous response is represented by

$$y_h(s) = 1 + \frac{m_2}{m_1}s + \frac{m_3}{m_1}s^2 + \frac{m_4}{m_1}s^3 + \cdots \tag{3.80}$$

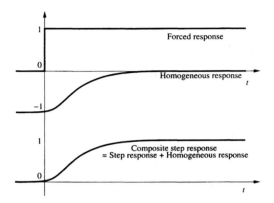

Figure 3.16.   The step response as a sum of forced and homogeneous responses [LAP98].

Interpreting this as a probability density function, the mean and variance are given by

$$\mu_1 \quad = \quad -\frac{m_2}{m_1} \tag{3.81}$$

$$\mu_2 \quad = \quad 2\left(\frac{m_3}{m_1}\right) - \left(\frac{m_2}{m_1}\right)^2 \tag{3.82}$$

Since the gamma distribution has a mean of $\frac{n}{\lambda}$ and a variance of $\frac{n}{\lambda^2}$, these values can be fitted as follows

$$n = \frac{\mu^2}{\mu_2} \text{ and } \lambda = \frac{\mu}{\mu_2} \tag{3.83}$$

The approximate step response is then given by

$$y(t) = 1 + m_1 g_{\lambda,n}(t) \tag{3.84}$$

and its median can be determined from a preconstructed two-dimensional lookup table that depends on n and λ. When the input is a saturated ramp, similar techniques as for PRIMO may be used to find the delay from a three-dimensional lookup table, parameterized by n, λ and the input transition time.

The WED metric.   The above methods inspired [LKA02] to use a Weibull distribution instead of the gamma distribution, yielding the advantage of requiring smaller lookup tables. The algorithm for the WEibull-based Delay (WED) metric is simple and is listed in Table 3.1(a); for details of the derivations, the reader is referred to [LKA02]. The computation uses the Γ function, which is has the property, $\Gamma(1 + x) = \Gamma(x)$ for $x > 1$; therefore, by storing the value of $Gamma(y), 1 \leq y \leq 2$ in a table, this property can be used to find $Gamma(x)$ for any arbitrary $x \geq 1$.

Algorithm for the WED metric
1. Calculate circuit moments m_1 and m_2.
2. Calculate $r = m_2/m_1^2$.
3. Use Table (b) below to find $\theta(r)$.
4. Use Table (c) below to find $\Gamma(1 + \theta)$.
5. Set $\beta = -m_1/\Gamma(1 + \theta)$.
6. Return a 50% delay value of $\beta(\ln 2)^\theta$.

(a)

r	$\log_{10}(r)$	θ
0.63096	-0.2	0.48837
0.79433	-0.1	0.76029
1.00000	0.0	1.00000
1.25892	0.1	1.22371
1.58489	0.2	1.43757
1.99526	0.3	1.64467
2.51189	0.4	1.84678
3.16228	0.5	2.04507
3.98107	0.6	2.24031
5.01187	0.7	2.43305
6.30957	0.8	2.62371
7.94328	0.9	2.81262
10.00000	1.0	3.00000
12.58925	1.1	3.18607
15.84893	1.2	3.37098

(b)

x	$\cdot \Gamma(x)$
1.0	1.00000
1.1	0.95135
1.2	0.91817
1.3	0.89747
1.4	0.88726
1.5	0.88623
1.6	0.89352
1.7	0.90864
1.8	0.93138
1.9	0.96176
2.0	1.00000

(c)

Table 3.1. (a) Pseudocode for the WED metric (b) Lookup tables for θ (c) Lookup table for the Γ function.

The D2M metric. Strictly speaking, D2M [ADK01] does not belong to this class of metrics that are derived from a probabilistic interpretation, but is also fast to compute. Historically, it preceded the WED metric, and the simple formula is related to it, and hence it is presented here for completeness.

$$\text{Delay}_{\text{D2M}} = \frac{m_1^2}{\sqrt{m_2}} \ln 2 \qquad (3.85)$$

This metric is purely empirical, although it is proven to always be less than the Elmore metric.

3.8 REALIZABLE CIRCUIT REDUCTION

While several model order reduction methods have been described in this chapter, most of them lead to fairly complicated models. The TICER (TIme Constant Equilibration Reducer) algorithm [She99] presents a simple and scalable RC-in RC-out method for reducing large RC networks by eliminating "slow" and "quick" nodes. A notable feature is that it preserves the Elmore delays through RC ladders when only "quick" internal nodes are eliminated.

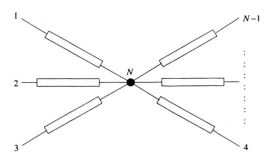

Figure 3.17. An representative node in an RC circuit, to be eliminated using TICER [She99].

The core of the method is based on an observation for a simple star network, such as that shown in Figure 3.17. The star has N nodes, with the peripheral nodes numbered from 1 to $N-1$, and the central node labeled N. Each branch from node N to node i, $i = 1 \cdots N - 1$, consists of a conductance g_{iN} and a capacitance c_{iN}[7].

If a step voltage is applied to one peripheral terminal i and the other $N - 2$ peripheral nodes are grounded, then the circuit may easily be analyzed, and the response at the central node N is found to be

$$h_{iN} = \frac{g_{iN}}{\gamma_N} + \left(\frac{c_{iN}}{\chi_N} - \frac{g_{iN}}{\gamma_N} \right) e^{-\frac{t}{\tau_N}} \qquad (3.86)$$

where

$$\gamma_N = \sum_{k=0}^{N-1} g_{kN}$$

$$\chi_N = \sum_{k=0}^{N-1} c_{kN}$$

$$\tau_N = \frac{\chi_N}{\gamma_N}$$

It is notable that regardless of which node $i \in \{1 \cdots N - 1\}$ is excited, the time constant τ_N remains the same, so that it is meaningful to associate a time constant with node N. Depending on whether this time constant is

"quick," "slow," or "normal," various approximations may be made. In particular, TICER presents a method for eliminating quick and slow nodes using a
reasonable approximation.

As usual, let us consider a circuit described by the MNA equations

$$(G + sC)\mathbf{X} = \mathbf{J} \tag{3.87}$$

We may rearrange these equations if necessary to ensure that the voltage of
the node N to be eliminated corresponds to the last element in $\mathbf{X}$. Therefore,
the above equation may be rewritten as

$$\begin{bmatrix} \tilde{Y} & y \\ y^T & (\gamma_N + s\chi_N) \end{bmatrix} \begin{bmatrix} \tilde{X} \\ v_N \end{bmatrix} = \begin{bmatrix} \tilde{J} \\ j_N \end{bmatrix} \tag{3.88}$$

If we eliminate v_N, we obtain

$$(\tilde{Y} - E)\tilde{X} = \tilde{J} - F \tag{3.89}$$

where

$$E_{ij} = \frac{(g_{iN} + sc_{iN})(g_{jN} + sc_{jN})}{\gamma_N + s\chi_N} \tag{3.90}$$

$$F_i = \left(\frac{g_{iN} + sc_{iN}}{\gamma_N + s\chi_N} \right) j_N \tag{3.91}$$

This may be used as a basis for node elimination as follows. For simplicity,
we will assume here that $j_N = 0$, although this is not essential; for the more
general case where $j_N \neq 0$, the reader is referred to [She99].

For quick nodes, $s\chi_N \ll \gamma_N$, and E_{ij} can be approximated as follows

$$E_{ij} \approx \frac{g_{iN}g_{jN}}{\gamma_N} \left(1 - \frac{s\chi_N}{\gamma_N} \right) + s\frac{g_{iN}c_{jN} + g_{jN}c_{iN}}{\gamma_N} \tag{3.92}$$

$$\approx \frac{g_{iN}g_{jN}}{\gamma_N} + s\frac{g_{iN}c_{jN} + g_{jN}c_{iN}}{\gamma_N} \tag{3.93}$$

A physical interpretation of this yields the following recipe for eliminating
quick nodes.

1. Remove all resistors and capacitors connected to node N.

2. Between all pairs of former neighbors i and j of N, insert the following
   new elements: a conductance $g_{iN}g_{jN}/\gamma_N$, and a capacitance $g_{iN}c_{jN} + g_{jN}c_{iN}/\gamma_N$.

For slow nodes, $s\chi_N \gg \gamma_N$, and E_{ij} can be approximated as follows

$$E_{ij} \approx \frac{g_{iN}g_{jN}}{\gamma_N} + s\frac{c_{iN}c_{jN}}{\chi_N} \tag{3.94}$$

The corresponding recipe for node elimination is as follows

1. Remove all resistors and capacitors connected to node N.

2. Between all pairs of former neighbors i and j of N, insert the conductance $g_{iN}g_{jN}/\gamma_N$ and the capacitance $c_{iN}c_{jN}/\chi_N$.

The basic TICER procedure described above is applicable to RC circuits. Subsequent work in [ACI03] has extended the algorithm to circuits with RLCK elements. An alternative approach based on the use of Y-Δ transforms is presented in [QC03], and is claimed to be significantly faster than TICER.

3.9 SUMMARY

This chapter has overviewed a set of techniques for reduced-order modeling of linear systems. Beginning with the simple Elmore delay metric, we have surveyed model order reduction methods, ranging from AWE to PRIMA, as well as fast delay metrics based on probability interpretations, and methods for building realizable circuits. In the next chapter, we will see how these techniques fit into a circuit-level timing analyzer.

Notes

1. It is possible to build causal delay metrics where the delay is guaranteed to be greater than or equal to zero. One possible way of doing this is to use the unity gain points in the DC transfer characteristic as crossing points for the input and output waveforms; however, such a metric will involve considerably more complex calculations and is therefore not generally used in practice.

2. An implicit assumption that is made when we speak of a dominant pole is that there is no nearby zero that can cancel out its effect.

3. The expression for the centroid is normally divided by the integral of $e'(t)$ from $t = 0$ to $t = \infty$; however, the equation below takes this simple form since $\int_0^\infty e'(t)dt = e(\infty) - e(0) = 1$.

4. The transfer function of a single-input single-output linear system is a ratio of the response at the output to the stimulus applied at the input [Kuo93].

5. The credit for this method is generally given to the French mathematician, Henri Padé (1863-1953) who provided the first thorough treatment of the topic.

6. The proof of this is simple and follows from the initial value theorem.

7. If no conductance or capacitance exists, then the value could simply be set to zero.

4 TIMING ANALYSIS FOR A COMBINATIONAL STAGE

4.1 INTRODUCTION

In Chapter 3, several techniques for the analysis of linear systems, such as interconnect systems, were presented. In this chapter, we will first consider the problem of analyzing the delay of a stage of combinational logic, consisting of a logic gate and the interconnect wires driven by it, and then extend it to handle coupled interconnect systems. This is a mixed linear/nonlinear system, since the interconnect consists of purely linear elements, while the logic gate models involve significant nonlinearities.

A simple way to address this problem is to approximate the nonlinear elements using linear models. In this case, the drain-to-source resistance of a transistor can be represented by a linear resistor. Under such a model, the theory from Chapter 3 may be used to solve the resulting linear system. However, the accuracy of this method will be limited since real-life transistor models show strong nonlinearities.

This chapter will present techniques for finding the delay of a stage of combinational logic, including both the nonlinear drivers and the linear interconnects. We will begin our presentation under the assumption that the wires show no capacitive coupling, and then extend the discussion to consider coupled interconnects.

4.2 IDENTIFYING A LOGIC STAGE

In general, a circuit may be specified in terms of a netlist that enumerates a set of circuit elements, such as transistors, resistors, capacitors and inductors. Each element is specified in terms of its electrical properties and the nodes to which its terminals are connected.

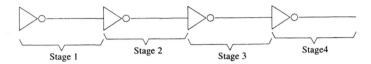

Figure 4.1. An example of an inverter chain, showing four stages of combinational logic.

In case of a full circuit simulation, one may write equations for the entire netlist and solve them, as in Chapter 2. However, for digital circuits, successive stages of combinational logic may be analyzed independently of each other, and the delay along a path may be computed simply as the sum of the delays of all of its combinational stages. The speed of the simulation can be greatly enhanced by taking advantage of this fact. Figure 4.1 shows a chain of inverters, each driving an interconnect wire. In this case, the four stages are identified, and the delay of the path may be taken to be the sum of the delays of these stages. In the ensuing discussion, we will refer to a component and the interconnect that it drives as a "logic stage."

Therefore, a first task that must be accomplished is to start from the input netlist, and to identify the combinational stages in the circuit. Roughly speaking, a combinational stage consists of a gate and the interconnect that it drives, but a stronger and more general definition is required. For example, the circuit shown in Figure 4.2 consists of an inverter driving a transmission gate. This would not conventionally be considered to be a single logic gate, but for the purposes of timing analysis, these are considered together since the behavior of the transmission gate and the inverter are very tightly coupled.

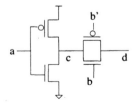

Figure 4.2.   An inverter driving a pass gate.

The key to finding reasonably decoupled stages is to observe that successive levels of logic are separated by gate nodes of transistors. Therefore, the basic unit that we must identify is a *channel-connected component* (henceforth referred to simply as a *component*). Each component corresponds to a set of transistors that are connected by drain and source nodes.

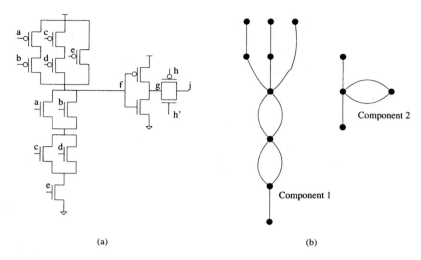

(a)   (b)

Figure 4.3.   (a) An example CMOS logic circuit (b) Its connection graph, showing its two channel-connected components.

More formally, the definition of a component can be given by the following construction. The circuit is represented as an undirected graph, $\mathcal{G}$, with a vertex for each circuit node and an edge between the drain and source nodes of each transistor. Next, the vertices corresponding to the ground and the supply (V_{DD}) nodes are split such that each of these vertices is incident on only one edge after splitting, resulting in a new graph, $\mathcal{G}'$. A component is then a set of transistors corresponding to the edges within a connected component of the graph, $\mathcal{G}'$. This process is illustrated by an example in Figure 4.3. The graph $\mathcal{G}'$ corresponding to the circuit in Figure 4.3(a) is shown in Figure 4.3(b), and it is easily seen that it has two strongly connected components, which correspond to the channel-connected components of the circuit.

The input nodes of a component consist of all of the gate nodes of transistors in the component, and any drain or source node of a transistor in the component that is also a primary input to the circuit. A component's output nodes include any drain or source node of a transistor in the component that are connected to a gate node of a transistor or to a primary output of the circuit.

On occasion, a component-based analysis may result in tremendously large components: for example, in case of a very large bus with tri-state drivers, or a barrel shifter. In such situations, it is more practical to partition the component into smaller segments. This is often done in an *ad hoc* way, and there is little literature that explores this issue in detail.

4.3   DELAY CALCULATION UNDER PURELY CAPACITIVE LOADS

Consider a logic gate driving an interconnect net to one or more sinks. If the entire net can be considered to be entirely capacitive, the problem of determin-

ing the delay of this logic stage reduces to that of finding the delay of the gate
for a specific capacitive load. If the gate is a cell from a library, its delay is typically precharacterized under various loads C_L and input transition times τ_{in}.
Typically, these methods are used to characterize the delay from each input pin
to the output, under the assumption that only one input switches at any time.
This is illustrated for the case of a falling output transition for a three-input
NAND gate in Figure 4.4. The delay from a to out here is computed using the
input excitations shown in Figure 4.4(b), where the inputs b and c are set to
logic 1 so as to propagate the input transition from a to the output. In practice,
the assumption of a single switching input is not entirely accurate, and the case
of simultaneously switching inputs (e.g., when inputs a and b switch together
in this example) can significantly impact the delay. Therefore, any model that
considers these will be more accurate [?S96, CGB01].

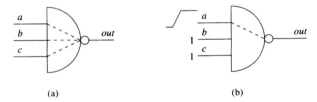

(a)   (b)

Figure 4.4. An illustration of the single switching input assumption that is widely used in
gate delay modeling.

4.3.1 Common delay metrics

Some of the commonly used delay characterization techniques under capacitive
loads include:

- A look-up table may be used, with each entry corresponding to the delay
  under different capacitive loads and input transition times. While this can
  achieve arbitrarily good accuracy provided the table has enough entries, in
  practice, it is seen to be memory-intensive.

- Traditional methods for characterizing a cell driving a load use an equation
  of the form
  $$k_1 C_L + k_2 \tag{4.1}$$
  where k_1 is a characterized slope and k_2 is an intrinsic delay. However, such
  an equation neglects the effect of the input transition time on the delay. One
  method that had often been applied is to multiply the delay of a gate to a
  step-input response (as calculated above) with a factor that depends on the
  input transition time [HJ87].

- The k-factor equations compact the table look-up by storing the delay D as
  a fitted function of the form
  $$(k_1 + k_2 C_L)\tau_{in} + k_3 C_L^3 + k_4 C_L + k_5 \tag{4.2}$$

- The nonlinear delay model (NLDM) from Synopsys uses characterization equations of the form

$$\alpha \tau_{in} + \beta C_L + \gamma \tau_{in} C_L + \delta \qquad (4.3)$$

- The scalable polynomial delay model (SPDM) from Synopsys uses a product of polynomials to fit the delay data. For example, for two parameters C_L and τ_{in}, a $P_m(C_L)Q_n(\tau_{in})$ fit is the product of an mth order polynomial in C_L with an nth order polynomial in τ_{in}, of the form

$$(a_0 + a_1 C_L + \cdots + a_m C_L^m) \cdot (b_0 + b_1 \tau_{in} + \cdots + b_n \tau_{in}^n) \qquad (4.4)$$

This approach is a generalization of the NLDM model, and has been found to be more compact.

- The above expressions perform a characterization for a single cell type, with fixed transistor sizes. Generalized posynomial models are more ambitious in that they attempt to find a unified expression for the delay over all allowable transistor widths, and will be presented in Section 8.7.1.

Similar equations are also used to characterize the output transition time.

Other equations have also been presented, based on analytic delay models. However, any equations based on the Shichman-Hodges model (the Spice Level 1 model that was used in an example in Section 2.4) is inherently inaccurate, since this model does not scale well to deep-submicron and nanometer technologies. Other models based on more complex semianalytic models such as the alpha-power model [SN90] and its variants scale a little better, but as technology moves into the nanometer domain and newer effects come into play, BSIM-like table lookup models are likely to be the only really accurate models, so that generating an analytic closed form will become harder, if not impossible.

Some delay models may have to factor in technology-specific issues. For instance, for partially depleted SOI technologies, the uncertainty in the body voltage adds another dimension to the delay estimation problem. The work in [SK01] provides a technique for overcoming this issue.

4.3.2 Finding the input logic assignments for the worst-case delay

A frequent problem that arises in timing analysis relates to determining the set of transistors that must be on during the worst-case delay calculations for a gate. One possible approach to this is to use the intuition that the Elmore delay is likely to have good fidelity (this was shown, albeit in a slightly different context, in [BKMR95]), and therefore can be used to determining the *identities* of these transistors. Once this set is determined, a more accurate model may be used for delay computation.

The most precise way of identifying the set is through a full enumeration; unfortunately, for a k input gate, this results in an exponential number of enumerations, which can be particularly severe for complex gates. The use of the Elmore fidelity argument results in heuristics (and intuition) that can significantly improve upon this. We will now present such a procedure.

The delay of a gate depends not only on the transistors that are on the path from the output to the supply node, but also on other off-path nodes whose parasitic capacitances must be charged or discharged in the process. The algorithm for finding the worst-case fall delay at an output node o of a gate is described below; the worst-case rise delay at o can be found in an analogous manner.

Consider a CMOS gate: we represent this by an undirected weighted graph, G, with an edge between the drain and source nodes of each transistor in the gate. The V_{dd} node and its incident edges are removed, since for a fall transition, the worst-case path will not involve this node or these edges. Edge weights are given by the resistance R_{on} of the corresponding transistor, which are inversely proportional to the transistor size; each node in the original circuit also has a capacitance associated with it, which is not directly represented in the graph.

The objective is to determine the set of edges that must be on to induce the largest Elmore delay in the component for a switching event on the transistor represented by edge e. It is reasonable to assume that this implies that there will only be one path from the output node to ground, and as a reasonable heuristic, we may assume that this is the path through e with the largest resistance[1]. We refer to this as the *largest resistive path* (LRP) through e, and it may be computed from the graph by finding the path of maximum weight between o and ground, which also passes through e.

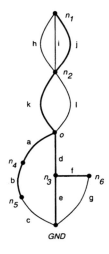

Figure 4.5.   Extending the LRP to a tree [SRVK93].

Since finding the LRP is equivalent to the longest path problem in a graph which is NP-complete, a heuristic must be used to perform this task. One such heuristic is provided in [SRVK93]. An alternative approach is to apply the Jouppi rules [Jou87a] to assign directions to transistors, which means that the longest path problem must now be solved on a directed graph; this is well known to be solvable in polynomial time [CLR90].

However, there may well be transistors that are not on the LRP that are on in the worst case. To motivate the intuition for this, consider the graph shown in Figure 4.5. Assume that the LRP between the output node o and ground through d has been found to be d, e. If, for example, transistor a is also on, then the capacitance that must be driven by the resistors on d and e is increased. Assuming the Elmore delay fidelity argument, the worst case delay scenario corresponds to the case when the downstream capacitances are chosen in such a way as to maximize the Elmore delay. Essentially, this implies that the edges should be chosen in the following order: first, maximize the downstream capacitance at the output node o by setting the appropriate transistors to "on," and then move along the LRP towards the ground node, each time setting transistors to "on" to maximizing the downstream capacitance at that node, subject to the assignments already made.

In this specific case, at node o, we may set transistors j, k, a and b to be "on." Next, moving to node n_3, we may set transistor i to be on. The worst case assignment then corresponds to the case where the darkened edges in the figure correspond to the on transistors. A more accurate calculator may now be used for the delay computation from the input of transistor d to the output o.

This procedure does not incorporate the effects of correlations between the inputs of a gate; one procedure that considers the case where the same signal may drive multiple pins of a gate is described in [DY96].

4.4 EFFECTIVE CAPACITANCE: DELAYS UNDER RC LOADS

4.4.1 Motivation

The lumped capacitance model for interconnects is only accurate when the driver resistance overwhelms the wire resistance; when the two are comparable, such a model could have significant errors. In particular, the phenomenon of "resistive shielding" causes the delay at the driver output (referred to as the *driving point*) to be equivalent to a situation where it drives a lumped load that is <u>less</u> than the total capacitance of the interconnect, as shown in Figure 4.6. In effect, the interconnect resistance shields a part of the total capacitance from the driving point. In this section, we will examine techniques that may be used to derive a value for the *effective capacitance* at the driving point. Such a capacitive model is particularly useful since it implies that cells may continue to be characterized in terms of a load capacitance as in Section 4.3 even in the domain where interconnect resistance is a dominant factor. The difference is that the load capacitance is no longer the total capacitance driven by the gate, but a smaller value corresponding to the effective capacitance.

As stated earlier, the parasitics that are associated with MOS transistors, particularly the transistor resistance, show significant nonlinearities, while the wire parasitics are linear. To take advantage of this, many timing analyzers process a logic stage in two steps:

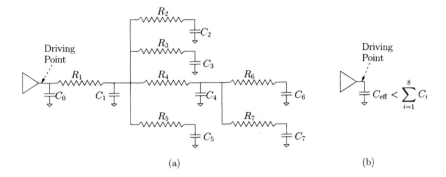

Figure 4.6. The notion of resistive shielding and the effective capacitance. A circuit consisting of a gate driving an RC tree is shown in (a), and in (b), the load is represented by an effective capacitance, C_{eff}. The value of C_{eff} is less than the total capacitance of the RC tree, $\sum_{i=1}^{7} C_i$ [QPP94].

- The waveform at the output of the logic gate, or the driving point, is first found using a composite model for the parasitics. This is a system that contains nonlinearities that stem from the transistor models, and this analysis must incorporate such effects.

- Next, this waveform at the driving point is used to drive the purely linear parasitic system, using linear system analysis techniques.

The primary advantage of such an approach is that it isolates the nonlinearities from the linearities, so that the efficient techniques from Chapter 3 may be applied in the second step. The problem of how to build a composite model for the parasitics through an "effective capacitance" approach will be addressed here.

4.4.2   The O'Brien-Savarino reduction

The O'Brien-Savarino reduction approach [OS89] reduces an arbitrary RC load to an equivalent RC-π model at the driving point, and is used as a starting point for finding C_{eff}. Consider a gate driving an RC line, as shown in Figure 4.7(a). An impulse voltage excitation $V(t) = \delta(t)$, corresponding to a unit voltage in the s domain, $V(s) = 1$, is applied at the root of the RC line, as shown in Figure 4.7(b) (or in general, an RC tree). One may apply the techniques of Section 3.5 to find the moments of the current, $I(s)$, through the voltage source. The moments of the admittance at the root can be computed in the form

$$Y(s) = \frac{I(s)}{V(s)} = y_0 + y_1 s + y_2 s^2 + \cdots \tag{4.5}$$

These are simply the moments of $I(s)$ when a unit voltage source in the s domain (i.e., $\delta(t)$ in the time domain) is applied.

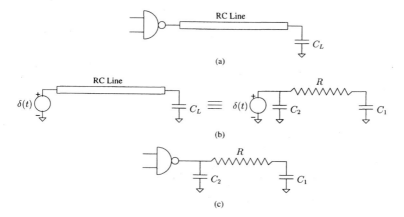

Figure 4.7. An illustration of the O'Brien-Savarino reduction on a circuit shown in (a). The interconnect is reduced to a π model, as shown in (b), which is then used to replace the load, as seen in (c).

This admittance is then represented by a reduced order model of the type shown in Figure 4.7(c)[2], whose admittance moments are computed as follows:

$$Y_{model} \;=\; sC_2 + \frac{sC_1}{1 + sRC_1} \tag{4.6}$$

$$\approx\; (C_1 + C_2)s + \sum_{i=2}^{\infty} \left[(-1)^{i-1} R^{i-1} C_1^i \right] s^i \tag{4.7}$$

To compute the three unknowns, R, C_1 and C_2, we may match these moments with those of the first three moments of $Y(s)$ to obtain a system of three equations in three variables. These are solved to yield:

$$C_1 \;=\; \frac{y_2^2}{y_3} \tag{4.8}$$

$$C_2 \;=\; y_1 - \frac{y_2^2}{y_3} \tag{4.9}$$

$$R \;=\; -\frac{y_3^2}{y_2^3} \tag{4.10}$$

An important consideration is that the circuit should be *realizable*, i.e., the resistance and capacitance values obtained above should all be nonnegative. It can be shown [KK00] that the first three admittance moments of general RC circuits satisfy

$$y_1 > 0, y_2 < 0, y_3 > 0 \tag{4.11}$$

$$y_1 y_3 - y_2^2 > 0 \tag{4.12}$$

From relations (4.8)–(4.12), it is easy to see that for an RC line, the reduction will always be realizable.

4.4.3 Effective capacitance computations

The use of the O'Brien-Savarino reduction to compute the gate delay still requires the calibration of the delay of each gate with respect to the four parameters, τ_{in}, R, C_1 and C_2, so that a four-dimensional look-up table must be created for the gate delay. With p points along each axis, this amounts to a table with p^4 entries, clearly worse than the p^2 entries required for a purely capacitive load. Even if curve-fitted formulæ were to be used, the complexity with four parameters would be much worse than with just two. The effective capacitance, C_{eff}, finds an equivalent capacitance that can be used to replace the π model, so that the delay characterization could continue to be performed over two parameters.

The essence of the C_{eff} calculations lies in creating a model that draws the same average current from a source as the RC-π model above, up to the 50% delay time point. In order to achieve this, the gate output must be modeled by an equivalent representation. Early efforts [QPP94] used a voltage source to model the gate output, but it was later found [DMP96] that a Thevenin model [NR00] with a voltage source and a resistor is more effective[3].

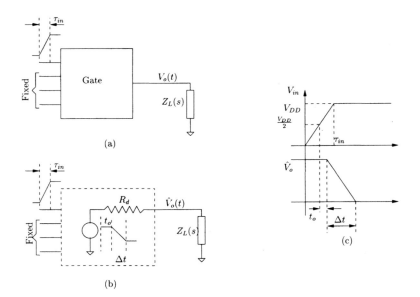

Figure 4.8. A Thevenin model for the driver for C_{eff} computations [DMP96]. (a) The original combinational stage consisting of a driver with one transitioning input, with a load $Z_L(s)$. (b) The Thevenin model. (c) The relation between the gate input and the Thevenin voltage source.

Figure 4.8(a) and (b) show gate driving a load, $Z_L(s)$, and its corresponding representation. Under the single switching input assumption, one input of the gate switches, and this is modeled by a saturated ramp with a rise time of τ_{in}, while the others are held steady. The Thevenin model consists of a resistance R_d and a voltage source $V_s(t)$ in the form of a saturated ramp that is parameterized by the variables t_0, the delay with respect to the 50% point of the input waveform, and Δt, the transition time, as illustrated in Figure 4.8(c). Therefore, characterizing the gate involves the determination of the values of R_d, t_0, and Δt.

In the subsequent discussion, we will refer to the $n\%$ point of a waveform as the time at which $n\%$ of the transition is complete. For a rising waveform, this is simply the time at which it reaches $n\%$ of V_{DD}, and for a falling waveform, this is the time when the signal is at $(100 - n)\%$ of V_{DD}.

The saturated ramp model for the Thevenin source can be written in terms of the difference of two time-shifted infinite ramps. For a falling output, such an equation would be

$$V_s(t, t_0, \Delta t) = \begin{cases} V_{DD} & t \leq t_0 \\ V_{DD} - r(t, t_0, \Delta t) & t_0 \leq t \leq t + \Delta t \\ V_{DD} - r(t, t_0, \Delta t) + r(t, t_0 + \Delta t, \Delta t) & t \geq t + \Delta t \end{cases}$$

(4.13)

where $r(t, \tau, \Delta t) = V_{DD} \left(\frac{t - \tau}{\Delta t} \right) u(t - \tau)$ is the ramp function and $u(t)$ is the unit step function. A similar equation can be written for the rising output transition.

When the load $Z_L(s)$ is a capacitor C_L, the response to this excitation is given by

$$\hat{V}_o(t, t_0, \Delta t) = \begin{cases} V_{DD} & t \leq t_0 \\ V_{DD} - y_0(t, t_0, \Delta t) & t_0 \leq t \leq t + \Delta t \\ V_{DD} - y_0(t, t_0, \Delta t) + y_0(t, t_0 + \Delta t, \Delta t) & t \geq t + \Delta t \end{cases}$$

(4.14)

where $y_0(.)$ is the response to an infinite ramp $r(t, t_0, \Delta t)$, and is given by

$$y_0(t, t_0, \Delta t) = V_{DD} \left[\frac{t - t_0}{\Delta t} - \frac{R_d C_L}{\Delta t} \left(1 - e^{-\frac{t - t_0}{R_d C_L}} \right) \right] \qquad (4.15)$$

Later, Equation (4.14) will be used to find the the 20% of 50% point of the $\hat{V}_o$ transition. Note that this involves solving an implicit nonlinear equation, and an iterative procedure, such as one based on the Newton-Raphson method, is used for this.

Finding R_d

The effective capacitance computation is found to be relatively insensitive to the value of R_d, and therefore, several simplifying assumptions may be used during its calibration. This procedure involves matching the delay of a resistor R_d driving a load capacitor C_L under the assumption of a step input voltage excitation to this RC structure. The insensitivity of the effective capacitance to R_d also enables flexibility in the choice of C_L: a reasonable value of C_L that is used is the largest capacitance that the cell is expected to drive.

When the gate drives a capacitive load under a step input, if the falling output waveform is assumed to be of the form

$$V_o(t) = V_0 e^{-\frac{t}{R_d C_L}} \tag{4.16}$$

then the value of R_d can be obtained from the 50% and 90% time points (denoted as t_{50} and t_{90}, respectively) as

$$R_d = \frac{t_{90}(C_L, \tau_{in}) - t_{50}(C_L, \tau_{in})}{C_L \ln 5} \tag{4.17}$$

The reason for choosing t_{50} and t_{90} rather than any other threshold is a result of empirical observations.

The values of t_{50} and t_{90} for a gate driving a purely capacitive load may easily be found using any of the techniques described in Section 4.3, such as the use of k-factor equations.

Finding t_0, C_{eff} and Δt

This procedure involves a set of iterations, whereby the average load current driven by the Thevenin model for the driver through an O'Brien-Savarino RC-π model, as shown in Figure 4.9(a), is equated to that through a capacitor C_{eff}, shown in Figure 4.9(b). Equivalently, this may be thought of as equating the total charge delivered to both circuits over a given time period, t_{av}. The value of t_{av} is typically set to Δt.

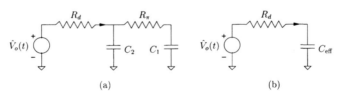

Figure 4.9. (a) The original combinational stage consisting of a driver with one transitioning input, with a load $Z_L(s)$. (b) The Thevenin model. (c) The relation between the gate input and the Thevenin voltage source [DMP96].

Since the value of the effective capacitance is unknown, this procedure involves iterations to compute the values of t_0, Δt and C_{eff}. Specifically, the iterations involve the following steps:

Step 1 The C_{eff} value is computed by solving the following equation, illustrated for a falling output transition:

$$I_\pi(\Delta t, t_{av}) = I_{C_{\text{eff}}}(\Delta t, t_{av}, C_{\text{eff}}) \tag{4.18}$$

As stated above, this equates the total charge delivered to the π load, I_π, with that sent to the C_{eff} load, $I_{C_{\text{eff}}}$. This computation presumes a value of Δt for the Thevenin source, which is taken from the previous iteration. The Newton-Raphson method is applied to iteratively solve this nonlinear equation to compute the value of C_{eff}.

The average currents on either side of Equation (4.18) are given by

$$I_\pi(\Delta t, t_{av}) = \left[A t_{av} + \frac{B}{p_1}\left(1 - e^{-p_1 t_{av}}\right) + \frac{D}{p_2}\left(1 - e^{-p_2 t_{av}}\right) \right] \frac{V_{DD}}{R_d t_{av} \Delta t}$$

$$I_{C_{\text{eff}}}(\Delta t, t_{av}, C_{\text{eff}}) = \left[R_d C_{\text{eff}} t_{av} - (R_d C_{\text{eff}})^2 \left(1 - e^{-\frac{t_{av}}{R_d C_{\text{eff}}}}\right) \right] \frac{V_{DD}}{R_d t_{av} \Delta t}$$

$$\tag{4.19}$$

where

$$A = \frac{z}{p_1 p_2}, \quad B = \frac{z - p_1}{p_1(p_1 - p_2)}, \quad D = \frac{z - p_2}{p_2(p_2 - p_1)}$$

and z, p_1 and p_2 represent the zero and two poles of the transfer function $\frac{\hat{V}_o(s) - V_s(s)}{V_s(s)}$ of Figure 4.9(a).

Step 2 Next, based on this computed value of C_{eff}, the values of t_0 and Δt for the Thevenin equivalent are updated using the following relations:

$$\hat{V}_o(t_{50}(C_{\text{eff}}), t_0, \Delta t) = 0.5 V_{dd} \tag{4.20}$$
$$\hat{V}_o(t_{20}(C_{\text{eff}}), t_0, \Delta t) = 0.8 V_{dd} \tag{4.21}$$

where t_{20} is the 20% point, i.e., the point at which 20% of the transition is complete; recall that for a falling transition, this corresponds to a voltage of $0.8 V_{DD}$. Here, the expression for $\hat{V}_o$ is provided by Equation (4.14), and the values of t_{50} and t_{20} on the left hand sides are based on the current value of C_{eff}, and can be found from characterized equations for the cell such as k-factor equations. These are two equations in two variables, and are solved using an iterative procedure that converges rapidly in practice.

Alternative approaches for C_{eff} computations are provided in [KM98, She02]. The former works with a fixed R_d and Δt and matches the delay instead of the total charge. The latter observes that most libraries are characterized for the 50% time, but not the 20% time, and proposes a method that avoids the requirement for an extra characterization for the 20% time.

4.4.4 Extension to RLC lines

For RLC lines, in principle, the effective capacitance approach suggested above may be applied. However, for such lines, the O'Brien-Savarino reduction is not guaranteed to be realizable, even for mildly inductive lines. In particular, it can be shown that when the admittance moments of the line are computed,

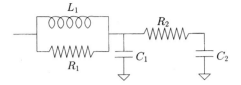

Figure 4.10.   A realizable reduced order model proposed in [KK00].

although $y_1 > 0$ and $y_2 < 0$, y_3 may take on any sign, so that the conditions in Equation (4.11) do not all hold true. Moreover, the validity of Equation (4.12) also depends on the inductance values in the circuit.

An example of an approach that overcomes this for RLC lines (with self-inductance only) is [KK00], which proposes a more general realizable model for RLC structures, shown in Figure 4.10. The unknowns here are the inductance L_1, the resistances R_1 and R_2, and the total capacitance C. These four parameters are matched against the first four moment terms of the impedance

$$Z(s) = \frac{1}{Y(s)} = \frac{z_{-1}}{s} + z_0 + z_1 s + z_2 s^2 + \cdots \qquad (4.22)$$

The total capacitance C is split into $C_1 = (1-k)C$ and $C_2 = (1+k)C$ where k is a parameter whose value is chosen so as to ensure realizability. Table 4.1 shows how k may be chosen for various combinations of the signs of z_1 and z_2. For various combinations of these signs, acceptable ranges of $x = \frac{k-1}{k+1}$ that guarantee realizability are provided, from which k may be determined.

z_1	z_2	$x = \frac{k-1}{k+1}$
> 0	< 0	Any positive real x; arbitrarily choose $x = 0$ ($k = 1$)
> 0	> 0	$x > \sqrt{z_2/(z_0^3 C^2)}$
< 0	< 0	$x > -z_1/(C z_0^2)$
< 0	> 0	$x > \max\left(-z_1/(C z_0^2), \sqrt{z_2/(z_0^3 C^2)}\right)$

Table 4.1.   Choosing the value of k to obtain a realizable RLC model.

Another approach for effective capacitance computation for RLC interconnects is presented in [ASB03]. This work shows that a single ramp is inadequate to model the output waveform in the presence of inductance, and proposes a piecewise linear waveform with two pieces that model the transition. The technique consists of the following steps. First, the breakpoint for the two pieces is determined. Next, separate C_{eff} values are found for each of the two pieces, and the output waveform is accordingly determined.

4.5   CAPACITIVE COUPLING EFFECTS

4.5.1   Introduction

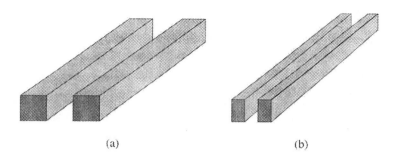

(a)   (b)

Figure 4.11.   An illustration of scaling trends as wire geometries shrink from a larger feature size in (a) to a smaller feature size in (b). The wires grow thinner and are closer spaces, but the aspect ratio becomes more skewed, leading to a larger proportion of coupling capacitance.

As feature sizes have shrunk with advancing technologies, wires have been brought closer to each other. While the width of a wire has correspondingly shrunk, wire heights have not scaled proportionally, as illustrated in Figure 4.11, since such scaling would increase wire resistances tremendously. The net effect of this has been to increase the capacitive coupling between wires, and this leads to two effects in terms of circuit performance:

Crosstalk noise   corresponds to noise bumps that are injected from an switching wire to an adjacent, nominally silent, wire. Several fast crosstalk noise metrics have been proposed in [Dev97, VCMS+99, KS01, CaPVS01, DBM03], and a detailed description of these is beyond the scope of this book.

Delay changes   occur, as compared to the uncoupled case, because of the additional coupling capacitances in the system.

The terminology that is frequently used in the context of crosstalk analysis is to label the wire being analyzed as the *victim*, and consider any wires that capacitively couple to it as *aggressors*.

In a full-chip analysis scenario that consists of a prohibitive number of aggressor/victim scenarios, it is vital to reduce the cases to be considered to a manageable number. The effect of an aggressor on a victim depends on a number of factors, and not every aggressor will inject an appreciable amount of noise into a victim. Pruning filters that identify unimportant coupling cases have been proposed in [LBB+00].

To illustrate the mechanisms that affect crosstalk, consider an example of two coupled lines of equal length. While this is a very simple case, the concepts illustrated here can be extended to understand crosstalk in the multiple line case. The equivalent segmented RC line with a Thevenin driver model is shown

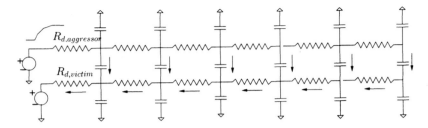

Figure 4.12. A circuit model for a pair of coupled lines. The aggressor switches from low to high, and injects $C\frac{dV}{dt}$ noise currents in the directions shown by the arrows. These currents flow through the line resistors and the driver (Thevenin) resistor to create a voltage bump along the line.

in Figure 4.12. A voltage change on the aggressor line creates a nonzero $\frac{dV}{dt}$ across the coupling capacitors, which injects currents into the victim. These currents may partially flow through the grounded capacitor on the victim, but mostly return through the resistances on the interconnect, through the transistors in the driver, to the supply node. Further analysis based on this mechanism points to the following factors that affect the magnitude of coupling noise.

- The spatial proximity of the lines determines the magnitude of the coupling capacitors, and influences the coupling noise.

- The temporal simultaneity of switching, as will be explained in Section 4.5.2, affects the $\frac{dV}{dt}$ across the capacitors, and is another factor.

- The noise at the far end of a line (near the receiver) is generally larger than that at the near end (near the driving point) since the injected noise current is dropped over a larger resistance.

- The victim impedance also affects the magnitude of the noise voltage; a larger impedance implies a larger drop.

- The switching rate of the aggressor influences the $\frac{dV}{dt}$ value across the coupling capacitor, so that a faster switching aggressor will induce a larger coupling noise.

Methods for reducing coupling effects include buffer insertion to reduce the distance to the far end of the line [ADQ98], adding additional spacing between wires, using shield lines in the form of power or ground lines, which maintain relatively stable voltage levels, sizing up the driver [BBA+03] and staggering repeaters on a bus [KMS99].

4.5.2   Miller capacitance models

A simple way to model crosstalk capacitances is to replace the coupling capacitor by a Miller capacitance to ground. To see how one can arrive at the value of this Miller capacitor, let us consider three cases, shown in Figure 4.13:

a. When a wire switches and is next to a nonswitching (silent) neighbor, the value of the Miller capacitance to ground equals the coupling capacitance. Although the silent wire will see a small noise bump on its voltage waveform, the time constant associated with this is large, so that the $\frac{dV}{dt}$ across the capacitor is substantially determined by the aggressor. Therefore, the the value of $\frac{dV}{dt}$ across the coupling capacitor and the Miller capacitor are identical, so that the $C\frac{dV}{dt}$ contribution of the former is well captured by the latter.

b. When the aggressor and the victim switch in the same direction, if the switching events track each other perfectly, then the value of $\frac{dV}{dt}$ across the coupling capacitor is zero. Using this intuition, the Miller capacitor is simply set to zero, which results in the same $C\frac{dV}{dt}$ current injection into the wire as from the coupling capacitor.

c. When the aggressor and the victim switch in opposite directions, if the changes on the two wires are symmetric, then $\frac{dV}{dt}$ will be twice that for case (a). This may be modeled by using a Miller capacitor value that is twice the value of the coupling capacitance.

The above cases are based on simplistic assumptions, and practically, for a coupling capacitance of C_c, instead of a $C_c/0/2C_c$ approximation, more realistic scenarios are "fudged" in using a $-C_c/0/3C_c$ approximation instead. This is not a completely unreasonable approximation, and has been shown to be exact under a specific coupling model in [CKK00].

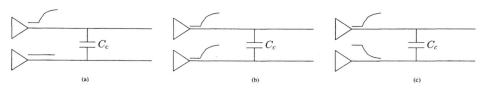

Figure 4.13. Three possible situations for a switching wire that is coupled to a neighbor: the adjacent wire (a) is silent (b) switches in the same direction (c) switches in the opposite direction. The analysis of line1 can be carried out "independently" of line2 by approximating the coupling capacitor by a grounded capacitance whose value is shown for each case. All other grounded capacitors on line1 remain unaffected by this transformation.

Although Figure 4.13 simplistically shows the lines coupled through a single coupling capacitor, the same reasoning is valid even for segmented coupled RC lines, such as that shown in Figure 3.8. In such a case, the coupling capacitor associated with each segment is converted to the corresponding Miller capacitance, depending on the direction in which the coupled wires switch.

4.5.3   Accurate coupling models and aggressor alignment

The use of the Miller capacitance is only a simple first-order approximation that can be used to extend the existing circuit analysis machinery for RC trees using Elmore delay computations, AWE, and the like. The primary advantage of this approximation is that it alters a coupled interconnect system to a set of uncoupled systems, each of which can be solved using the methods described earlier.

A more accurate picture of reality is illustrated for a pair of simultaneously switching coupled lines in Figure 4.14. The line of interest is the victim line, and the aggressor line influences it through a set of coupling capacitances that are not explicitly shown in the figure.

There are two sources of noise on a wire

- A noise bump at the input to a gate may propagate to its output, typically in an attenuated form. Extremely fast noise transitions at a gate input may not be sent through at all, since a gate acts as a low pass filter.

- The voltage on the aggressor line results in the injection of a noise voltage pulse on node v of the victim, as shown in Figure 4.14(b)[4].

Processing the effects of the first of these components exactly requires a topological traversal of the entire circuit to trace the noise propagation [SNR99], since the combinational stage at the input to a given stage must be processed prior to the current stage. To avoid the complications associated with such scenarios, which may result in cyclical dependencies when a topologically earlier stage is coupled to a topologically later stage, one may often impose a constraint that limits the propagated noise bump to be lower than a peak value and pessimistically assume the propagated noise to be at this peak value on each net. This yields the advantage that each net can be processed independently.

These two components add up to contribute to a noise voltage waveform at node v on the victim. Since the interconnect net, when driven by a Thevenin equivalent for the driver, is a linear system, the composite waveform at v can be obtained by adding the noise-free waveform, shown by the uppermost graph in Figure 4.14(c), with this noise bump. The exact time at which the noise is injected depends on the transition time of the aggressor, relative to that of the victim; this is referred to as the *aggressor alignment*. The second and third graphs in Figure 4.14(c) show the noise bump waveforms corresponding to two different aggressor alignments.

Note that the noise-free waveform, $V_{v,noisefree}$, has a delay of $d_{noisefree}$ in the figure, while the two different aggressor alignments have delays of d_1 and d_2, respectively, and the latter corresponds to the worst-case delay at v over all alignments. The task of delay computation in a coupled system involves the determination of aggressor alignments that result in the worst-case delay(s) at the node(s) of interest.

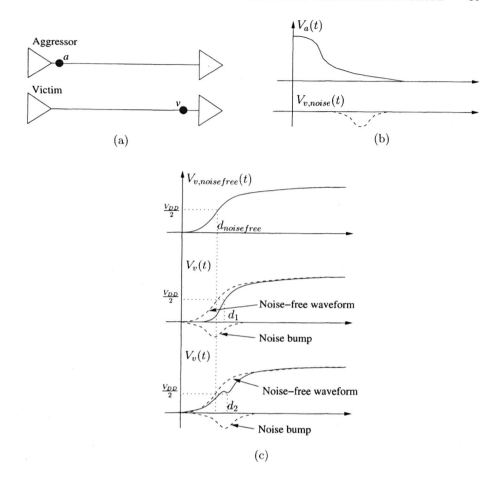

Figure 4.14. (a) A pair of coupled lines (to avoid cluttering the figure, the coupling capacitors are not shown). (b) The waveform at the driving point of the aggressor, and the corresponding noise voltage waveform, $V_{v,noise}(t)$, injected at node v of the victim. (c) The voltage at v can be obtained by adding the noise-free waveform, shown in the uppermost graph, with the noise pulse. The second and third plots show the superposition for different injection times for the noise pulse, corresponding to different aggressor alignments.

4.5.4 The concept of timing windows

The role of the coupling capacitances is greatly dependent on the relative switching times of the nets. Using the models of Section 4.5.2, depending on whether a neighboring wire is silent, or switches in the same or the opposite direction, a Miller capacitance of C_c, 0 or $2C_c$ (or some variant thereof) may be used to substitute the coupling capacitor. While this is more approximate than the analysis in Section 4.5.3, it is adequate to explain the concept of timing windows that is introduced in this section.

The use of a Miller capacitor transforms a coupled RC network into a set of uncoupled RC trees, and the values of the capacitors, and hence the delays, in these trees depend on the switching characteristics of the neighboring wires. Therein lies the complexity of the timing calculation procedure: the delays depend on the Miller capacitance values, which in turn depend on the switching times that are determined by the delays. This cyclic relationship makes the task of timing analysis in coupled RC networks nontrivial.

In general, one can define a timing window as an interval (or most generally, a union of intervals) in time during which a node in a circuit may switch. Given the timing windows for the aggressors and the victims at the driver inputs of a set of coupled lines, it is possible to determine the timing windows at the receivers driven by these drivers.

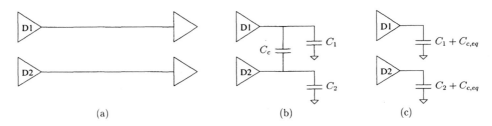

Figure 4.15.   An example to illustrate the idea of cyclic dependencies in computing timing windows. (a) shows a pair of coupled lines, and the equivalent lumped capacitance model is shown in (b). Based on the switching directions of the wires, the coupling capacitance on each line can be represented as a lumped capacitance to ground, $C_{c,eq}$; however, determining its values involves the resolution of a cyclic dependency.

As an example, consider two wires that are laid out adjacent to each other, as shown in Figure 4.15(a). For purposes of illustration, let us use the simplistic model in Figure 4.15(b), where the capacitances to ground as well as the coupling capacitance are all lumped; the analysis presented here will hold even for more complex distributed or segmented wire models. If the input signals to driver of the two wires switch between times $[T_{min,in1}, T_{max,in1}]$ and $[T_{min,in2}, T_{max,in2}]$, respectively, and if the delays required to propagate the signal along the wires are in the range $[d_{1,min}, d_{1,max}]$ and $[d_{2,min}, d_{2,max}]$, respectively, then the intervals during which the lines switch are $[T_{min,1}, T_{max,1}]$ and $[T_{min,2}, T_{max,2}]$, respectively, where $T_{min,1} = T_{min,in1} + d_{1,min}$, $T_{max,1} = T_{max,in1} + d_{1,max}$, $T_{min,2} = T_{min,in2} + d_{2,min}$, and $T_{max,2} = T_{max,in2} + d_{2,max}$. Therefore, the following relationship holds between the switching times and the equivalent coupling capacitance, $C_{c,eq}$.

The first line of Table 4.2 corresponds to the case where the switching intervals for the two lines overlap, and the value of $C_{c,eq}$ is accordingly chosen to be either 0 or $2C_c$, depending on whether the signals switch in the same direction,

Table 4.2.   Variation of $C_{c,eq}$ with switching time.

Interval	$C_{c,eq}$
$[\max\{T_{min,1}, T_{min,2}\}, \min\{T_{max,1}, T_{max,2}\}]$	0 or $2C_c$
$[\min\{T_{min,1}, T_{min,2}\}, \max\{T_{min,1}, T_{min,2}\}]$	C_c
$[\min\{T_{max,1}, T_{max,2}\}, \max\{T_{max,1}, T_{max,2}\}]$	C_c

or in opposite directions[5]. The second and third lines correspond to the cases where there is no overlap in the switching windows.

Since $d_{1,min}$, $d_{1,max}$, $d_{2,min}$ and $d_{2,max}$ are all dependent on the value of $C_{c,eq}$ (under, for example, a k-factor model), the values of $T_{min,1}$, $T_{max,1}$, $T_{min,2}$ and $T_{max,2}$ also depend on $C_{c,eq}$. However, an examination of this table shows that $C_{c,eq}$ is itself dependent on the values of $T_{min,i}$ and $T_{max,i}$, $i = 1, 2$.

To break this cyclic dependency, iterative approaches are often employed. The technique proposed in [Sap99, Sap00] provides a method that determines worst-case bounds on the number of iterations required for convergence. Other significant efforts in this direction include [Zho03], which shows the problem formulation in the context of a solution on a lattice, [TB03] that attempts to remove the need for iterations, and [ARP00], which employs this method in the context of a full static timing analysis for a circuit with multiple stages of logic.

In developing the concept of timing windows, three factors come into play:

Spatial neighborhood is clearly vital, since adjacency between nets is required for any capacitive coupling.

Temporal aspects come into play since the timing window, described in terms of the earliest and latest allowable switching times, must overlap in order to excite a simultaneous switching scenario corresponding to the maximum or minimum possible delay. If such overlap is not possible, then the maximum and minimum delay computations must be suitably adjusted.

Functional issues correspond to Boolean relationships between the signals on adjacent wires, and are also important. Even though the sum of the delays on a path may predict temporal adjacency, the Boolean relationships in the circuit may not permit the signals to switch simultaneously. The importance of this lies in the fact that timing windows are often computed rapidly using a static timing analyzer (which will be described in detail in Chapter 5), which is blind to the logical relationships in a circuit, and simply sums up the delays of gates on a path to find the maximum (or minimum) delay. Techniques such as [KSV96, CK99, ABP01] have been proposed to address the problem of integrating temporal and functional dependencies.

Therefore, timing window computations are vital to determining the effects of crosstalk on timing. Most approaches define timing windows in terms of an

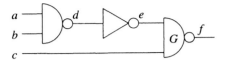

Figure 4.16.   A circuit example to illustrate the discontinuous nature of switching windows.

earliest and latest switching time, and assume that a switching event is permissible at any intermediate time. In practice, this is a pessimistic argument, and the true switching windows could be a union of disjoint windows. For instance, consider the example shown in Figure 4.16, where a, b, and c are primary inputs that switch at time 0. Assuming that the delay of each gate is 1, the minimum and maximum delays at nodes c and e, the inputs to gate G, can easily be seen to be 0 and 2 units, respectively. Accordingly, adding to this the unit delay of gate G, the switching window at f can be set to the interval $[1, 3]$. However, in reality, a switching event at f can only occur at this node at time 1, due to path $c - f$, or at time 3, due to path $a - d - e - f$ or $b - d - e - f$, and not at any arbitrary time in this interval (e.g., at time 2). Therefore, the choice of the <u>entire</u> interval $[1, 3]$ may enable some nonexistent simultaneous switching events, which will result in pessimistic calculations for both the maximum and minimum delays at the output of G.

4.5.5   Extending the effective capacitance method to coupled RC lines

We will illustrate the concept of C_{eff} computations from [DP97] on a pair of RC coupled lines, shown in Figure 4.17, but the idea can be extended to RC coupled trees with an arbitrary number of coupled nets. In the two-line case considered here, the first step is to replace each driver by a Thevenin model, as in Section 4.4.3; in this explanation, a Norton equivalent is used instead. This is characterized by the parameters t_{01} and Δt_1 for the first line and t_{02} and Δt_2 for the second, and these must be calculated, in addition to the resistors R_{d1} and R_{d2}.

Two effective capacitances, $C_{\text{eff},1}$ and $C_{\text{eff},2}$, one for each driver, and the four Thevenin parameters for the two driver models are computed as follows:

- An initial guess for the $C_{\text{eff},i}, i = 1, 2$ values is chosen.

- Considering this system as a two-port, the linear system is solved to find the voltage at each of the driving point nodes. Effectively, this method is identical to that considered for the RC line case, except for the addition of an extra current source in the load driven by each Thevenin equivalent.

- Next, for each line, the average voltage at the driving point for the coupled load is equated with that injected into the effective capacitance load. In

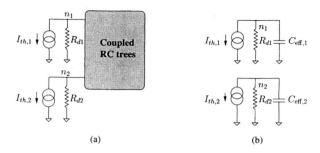

Figure 4.17. (a) A coupled RC network driven by the Norton equivalent for its driver (note that this can easily be translated to/from the Thevenin equivalent). (b) The equivalent effective capacitance model.

other words, we set

$$\int_{t_{01}}^{t_{01}+\Delta t_1} v_{actual,1}(t)dt = \int_{t_{01}}^{t_{01}+\Delta t_1} v_{C_{\text{eff},1}}(t)dt \qquad (4.23)$$

$$\int_{t_{02}}^{t_{02}+\Delta t_2} v_{actual,2}(t)dt = \int_{t_{02}}^{t_{02}+\Delta t_2} v_{C_{\text{eff},2}}(t)dt \qquad (4.24)$$

where t_{0i} and Δt_i correspond to the Thevenin parameters for driver i, as defined in Section 4.4.3. As before, this expression is used to update the value of C_{eff}. This can be shown to be equivalent to the average current (or total charge) formulation used earlier.

- The next task is to update the value of $t_{01}, t_{02}, \Delta t_1$ and Δt_2. This is achieved, as before, by using the driver models at the computed value of $C_{\text{eff},i}, i = 1, 2$ to match the 20% and 50% points on each line. These four equations here (two for each line) are essentially the counterparts of Equations (4.20) and (4.21) for the single line case. Together with the two equations (4.23) and (4.24) above, we obtain six equations that are solved iteratively to find the six unknown variables.

This procedure can be generalized to the n line case, where the number of variables is $3n$, corresponding to an effective capacitance variable and two Thevenin driver variables for each of the n lines. The $3n$ equations arise from a charge equation, a 20% matching equation and a 50% matching equation at each of the n driving points, and these can be solved in a manner similar to that outlined for the two-line case above.

4.5.6   Finding the worst-case aggressor alignment

As shown in Section 4.5.3, the precise aggressor alignment affects the delay of the waveform. We now explain how the technique described in the previous section can be extended to find the worst-case aggressor alignment, i.e., the alignment that results in the worst-case delay.

Consider a victim with a rising waveform that is capacitively coupled to several aggressors, each with a falling waveform, as illustrated in Figure 4.18(a). At a node of interest v on the victim, each aggressor a_i will contribute a noise waveform, $V_{a_i,v}(t)$. The technique in [GARP98] uses a procedure similar to waveform relaxation [LRSV83] to arrive at this solution. Each line is solved separately using a Thevenin model for each driver, and its contribution $V_{a_i,v}(t)$ at the victim node is computed. Since this system uses a Thevenin equivalent for each driver, it is purely linear, and the noise waveform can be found by summing up the individual contributions[6] $V_{a_i,v}(t)$. The worst-case noise peak, V_N, is achieved when the noise peaks of all of the aggressors are perfectly aligned, as shown in Figure 4.18(b)[7]. To determine the worst-case delay, this composite noise pulse is aligned with the noise-free victim waveform to create a worst-case scenario. This occurs when the noise peak is aligned with the time at which the noise-free victim is at voltage $V_{DD} + V_N$. Since the noise-free waveform rises monotonically, the sum of the two waveforms has its minimum at this time point, as shown in Figure 4.18(c).

Therefore, the process of determining the worst-case 50% delay at v for a rising victim waveform consists of the following steps:

1. The noise-free waveform is determined, as are the individual noise waveforms from each aggressor.

2. The peaks of the noise waveforms are all aligned to find V_N, the noise waveform peak; note that V_N is negative since all of the aggressors sport falling waveforms.

3. The noise-free waveform, $V_{noisefree}$, is now aligned with the noise signal in such a way that the time at which the latter reaches V_N corresponds to the time when $V_{noisefree}$ is at $V_{DD} + V_N$.

An enhancement of this approach was proposed in [BSO03]. Firstly, it is observed that the aggressor alignment strategy above uses the artifact of the 50% delay definition, which may not be very meaningful in practice. Specifically, when a noisy waveform is at the input to the victim receiver, the output waveform is shown in Figure 4.19. In this case, the pulse propagates to the output when the transient has already settled, and thus, the noise pulse does not appreciably change the waveform at the victim output. Consequently, an improved definition of the delay is necessary, whereby the worst-case aggressor alignment is a function of the gate type, size, pmos/nmos ratio, and output load on the victim receiver. An inexpensive precharacterization using a linear fit was proposed to capture these effects in this work.

Secondly, [BSO03] noted that the Thevenin resistance that is valid over the entire transition between 0 and V_{DD} is not identical to the resistance that is seen for the noise transition, which has a much smaller range. Therefore, it developed the concept of a *transient holding resistance* to capture this effect,

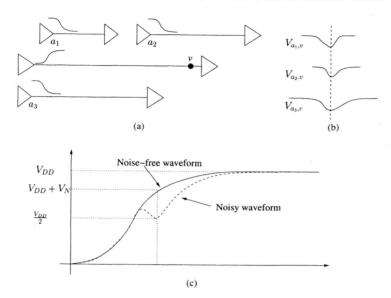

Figure 4.18.   (a) An example of a victim line with three aggressors, a_1, a_2 and a_3. (b) The noise pulses induced by the three aggressors are aligned to coincide in the worst case, and the composite noise peak, V_N, is the sum of all of these noise peaks. (c) The composite noise pulse, which is the sum of the pulses in (b), is added to the noise-free waveform to find the latest 50% crossing point for node v.

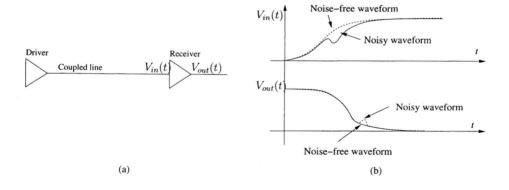

Figure 4.19.   (a) A net that is coupled to other nets (the coupling is not shown in the figure), with a driver and a receiver at the far end. (b) The noise-free and noisy waveforms at the input at output of the receiver, when the aggressors are aligned so as to find the latest 50% delay point at the receiver input, showing that the latest 50% crossing point at the receiver input does not necessarily correspond to the worst-case delay scenario [BSO03].

and developed a modified effective capacitance computation algorithm for computing its value.

In addition to these efforts, several other notable contributions have been made. TETA [ADP02] uses a simulation-based environment to avoid the inaccuracies of effective capacitance calculation. Its core is a fast nonlinear driver simulations using table-based transistor models and the parallel chord method for rapid model evaluation, and it employs a relaxation-based framework for the fast solution of a set of coupled lines. The approach in [CMS01] avoids the relaxation iterations and uses a direct technique to handle nonlinearities in the circuit.

4.6 SUMMARY

In this chapter, we have seen how the delay of a single combinational stage, with many coupled interconnect elements, can be found. A stage may consist of linear interconnect elements and nonlinear elements related to MOS devices. Efficient techniques for taking advantage of the linearities while taking the nonlinearities into account have been presented, including the notion of an effective capacitance. This provides the basis for circuit-level analysis, where the delays of combinational stages are added up to systematically determine the worst-case delay of a combinational circuit.

Notes

1. This assumption is valid when the load capacitances at internal nodes are small enough, as compared to the capacitance at the gate output. This is often the case in CMOS circuits.

2. The original paper [OS89] studied reductions to a single capacitor, a single RC and the π model described here; the latter has become the most widely used.

3. Alternatively, the Norton equivalent corresponding to the Thevenin model may be used.

4. In the figure, the voltage on the aggressor line is simplistically shown to be monotonic. In reality, it may also be affected by a noise bump. In any case, it will result in the injection of a noise voltage pulse that is similar in nature to that shown here.

5. Further subtlety can be added by considering rise and fall times independently, and the procedure can be suitably modified.

6. Note that this is a nonlinear system, and strictly speaking, superposition is not permitted. However, the nonlinear iterations are used to determine the parameters of the Thevenin model, and for small noise perturbations about this, a linear approximation is possible, which permits the use of superposition. If this addition is performed using the original driver models, it is seen to underestimate the noise on occasion.

7. Note that this does not mean that the input waveforms to the aggressors are aligned!

5 TIMING ANALYSIS FOR COMBINATIONAL CIRCUITS

5.1 INTRODUCTION

The methods described in Chapter 4 can be employed to find the delay of a single stage of combinational logic. A larger combinational circuit consists of several such stages, and the next logical step is to extend these methods for circuit-level delay calculation. This chapter will present methods that compute the delay of a combinational logic block in a computationally efficient manner.

5.2 REPRESENTATION OF COMBINATIONAL AND SEQUENTIAL CIRCUITS

A combinational logic circuit may be represented as a timing graph $G = (V, E)$, where the elements of V, the vertex set, are the logic gates in the circuit and the primary inputs and outputs of the circuit. Strictly speaking, this discussion should work with channel-connected components instead, but we will press on with the term "gate," with the understanding that it is considered to be equivalent to a component in this context.

A pair of vertices, u and $v \in G$, are connected by a directed edge $e(u, v) \in E$ if there is a connection from the output of the element represented by vertex u to the input of the element represented by vertex v. A simple logic circuit and its corresponding graph are illustrated in Figure 5.1(a) and (b), respectively.

A simple transform that converts the graph into one that has a single source s and a single sink t is often useful. In the event that all primary inputs are connected to flip-flops and transition at the same time, edges are added from the node s to each primary input, and from each primary output to the sink t. The case where the primary inputs arrive at different times $a_i \geq 0$ can be handled with a minor transformation to this graph[1], adding a dummy node with a delay of a_i along each edge from s to the primary input i.

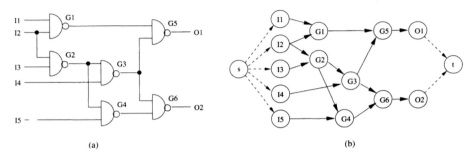

(a)   (b)

Figure 5.1.   (a) An example combinational circuit, and (b) its timing graph.

The above construction generally results in a combinational circuit that is represented as a directed acyclic graph (DAG). This is because combinational circuits, as conventionally described, do not have any cycles. However, many actual implementations of combinational circuits are not acyclic, and some literature on cyclic combinational circuits has been published in [Mal95, RB03, Edw03]. Many timing analyzers handle loops in combinational circuits by breaking them either in an *ad hoc* manner, or according to a heuristic, so that the circuit to be analyzed may be represented as a DAG.

A sequential circuit that consists both of combinational elements and sequential elements (flip-flops and latches) may be represented as a set of combinational blocks that lie between latches, and a timing graph may be constructed for each of these blocks. For any such block, the sequential elements or circuit inputs that fanout to a gate in the block constitute its primary inputs; similarly, the sequential elements or circuit outputs for which a fanin gate belongs to the block together represent its primary outputs. It is rather easy to find these blocks: to begin with, we construct a graph in which each vertex corresponds to a combinational element, an undirected edge is drawn between a combinational element and the combinational elements that it fans out to, and sequential elements are left unrepresented (this is substantially similar to the directed graph described above for a combinational circuit, except that it is undirected). The connected components of this graph correspond to the combinational blocks in the circuit.

The computation of the delay of such a combinational block is an important step in timing analysis. For an edge-triggered circuit, the signals at the primary inputs of such a block make a transition at exactly the same time, and the clock period must satisfy the constraint that it is no smaller than the maximum delay of any path through the logic block, plus the setup time for a flip-flop at the primary output at the end of that path. Therefore, finding the maximum delay through a combinational block is a vital subtask, and we will address this issue in this section. Alternatively, a combinational block that is to be treated as a black box must be characterized in terms of its delays from each input to each output; this is a vital problem in, for example, when the analysis involves hierarchically defined blocks with such timing abstractions. For level-clocked circuits, the timing relations are more complex, but the machinery developed in this section will nevertheless be useful. A detailed discussion on the analysis of sequential circuits based on combinational block delays will be presented in Chapter 7.

5.2.1 Delay calculation for a combinational logic block

In this section, we will present techniques that are used for the static timing analysis of digital combinational circuits. The word "static" alludes to the fact that this timing analysis is carried out in an input-independent manner, and purports to find the worst-case delay of the circuit over all possible input combinations. The computational efficiency of such an approach has resulted in its widespread use, even though it has some limitations that we will describe in Section 5.3.

A method that is commonly referred to as PERT (Program Evaluation and Review Technique) is popularly used in static timing analysis. In fact, this is a misnomer, and the so-called PERT method discussed in most of the literature on timing analysis refers to the CPM (Critical Path Method) that is widely used in project management[2]. The CPM procedure is now an integral part of most fast algorithms for circuit delay calculation, and in this book we will endeavor to use the correct terminology.

Before proceeding, it is worth pointing out that while CPM-based methods are the dominantly in use today, other methods for traversing circuit graphs have been used by various timing analyzers: for example, depth-first techniques have been presented in [Jou87b].

The algorithm, applied to a timing graph $G = (V, E)$, can be summarized by the pseudocode shown in Figure 5.2. The procedure is best illustrated by means of a simple example. Consider the circuit in Figure 5.3, which shows an interconnection of blocks. Each of these blocks could be as simple as a logic gate or could be a more complex combinational block, and is characterized by the delay from each input pin to each output pin. For simplicity, this example will assume that for each block, the delay from any input to the output is identical. Moreover, we will assume that each block is an inverting logic gate such as a NAND or a NOR, as shown by the "bubble" at the output. The two numbers, d_r/d_f, inside each gate represent the delay corresponding to the delay of the output rising transition, d_r, and that of the output fall transition, d_f, respectively. We assume that all primary inputs are available at time zero, so that the numbers "0/0" against each primary input represent the worst-case rise and fall arrival times, respectively, at each of these nodes. The critical path method proceeds from the primary inputs to the primary outputs in topological

```
Algorithm CRITICAL_PATH_METHOD
Q = ∅;
for all vertices i ∈ V
    n_visited_inputs[i] = 0;
/* Add a vertex to the tail of Q if all inputs are ready */
for all primary inputs i
    /* Fanout gates of i */
    for all vertices j such that (i → j) ∈ E
        if (++n_visited_inputs[j] == n_inputs[j]) addQ(j,Q);
while (Q ≠ ∅) {
    g = top(Q);
    remove(g,Q);
    compute_delay[g]
    /* Fanout gates of g */
    for all vertices k such that (g → k) ∈ E
        if (++n_visited_inputs[k] == n_inputs[k]) addQ(k,Q);
}
```

Figure 5.2.   Pseudocode for the critical path method (CPM).

order, computing the worst-case rise and fall arrival times at each intermediate node, and eventually at the output(s) of a circuit.

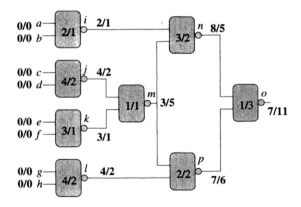

Figure 5.3.   An example illustrating the application of the critical path method (CPM) on a circuit with inverting gates. The numbers within the gates correspond to the rise delay/fall delay of the block, and the italicized bold numbers at each block output represent the rise/fall arrival times at that point. The primary inputs are assumed to have arrival times of zero, as shown.

A block is said to be *ready for processing* when the signal arrival time information is available for all of its inputs; in other words, when the number of processed inputs of a gate g, n_visited_inputs[g], equals the number of inputs of the gate, n_inputs[g]. Notationally, we will refer to each block by the symbol for its output node. Initially, since the signal arrival times are known only at the primary inputs, only those blocks that are fed solely by primary inputs are ready for processing. In the example, these correspond to the gates i, j, k and l. These are placed in a queue Q using the function addQ, and are processed in the order in which they appear in the queue.

In the iterative process, the block at the head of the queue Q is taken off the queue and scheduled for processing. Each processing step consists of

- Finding the latest arriving input to the block that triggers the output transition (this involves finding the maximum of all worst-case arrival times of inputs to the block), and then adding the delay of the block to the latest arriving input time, to obtain the worst-case arrival time at the output. This is represented by function compute_delay in the pseudocode.

- Checking all of the block that the current block fans out to, to find out whether they are ready for processing. If so, the block is added to the tail of the queue using function addQ.

The iterations end when the queue is empty. In the example, the algorithm is executed as follows:

Step 1 In the initial step gates i, j, k and l are placed on the queue since the input arrival times at all of their inputs are available.

Step 2 Gate i, at the head of the queue, is scheduled. Since the inputs transition at time 0, and the rise and fall delays are, respectively, 2 and 1 units, the rise and fall arrival times at the output are computed as 0+2=2 and 0+1=1, respectively. After processing i, no new blocks can be added to the queue.

Step 3 Gate j is scheduled, and the rise and fall arrival times are similarly found to be 4 and 2, respectively. Again, no additional elements can be placed in the queue.

Step 4 Gate k is processed, and its output rise and fall arrival times are computed as 3 and 1, respectively. After this computation, we see that all arrival times at the input to gate m have been determined. Therefore, it is deemed ready for processing, and is added to the tail of the queue.

Step 5 Gate l is now scheduled, and the rise and fall arrival times are similarly found to be 4 and 2, respectively, and no additional elements can be placed in the queue.

Step 6 Gate m, which is at the head of the queue, is scheduled. Since this is an inverting gate, the output falling transition is caused by the latest input rising transition, which occurs at time $\max(4, 3) = 4$. As a consequence, the fall arrival time at m is given by $\max(4, 3) + 1 = 5$. Similarly, the rise arrival time at m is $\max(2,1) + 1 = 3$. At the end of this step, both n and p are ready for processing and are added to the queue.

Step 7 Gate n is scheduled, and its rise and fall arrival times are calculated, respectively, as $\max(1,5) + 3 = 8$ and $\max(2,3) + 2 = 5$.

Step 8 Gate p is now processed, and its rise and fall arrival times are found to be $\max(5,2) + 2 = 7$ and $\max(3,4) + 2 = 6$, respectively. This sets the stage for adding gate o to the queue.

Step 9 Gate o is scheduled, and its rise and fall arrival times are $\max(5,6) + 1 = 7$ and $\max(8,7) + 3 = 11$, respectively. The queue is now empty and the algorithm terminates.

The worst-case delay for the entire block is therefore $\max(7,11) = 11$ units.

5.2.2   Critical paths, required times, and slacks

The *critical path*, defined as the path between an input and an output with the maximum delay, can now easily be found by using a traceback method. We begin with the block whose output is the primary output with the latest arrival time: this is the last block on the critical path. Next, the latest arriving input to this block is identified, and the block that causes this transition is the preceding block on the critical path. The process is repeated recursively until a primary input is reached.

In the example, we begin with Gate o at the output, whose falling transition corresponds to the maximum delay. This transition is caused by the rising transition at the output of gate n, which must therefore precede o on the critical path. Similarly, the transition at n is effected by the faling transition at the output of m, and so on. By continuing this process, the critical path from the input to the output is identified as being caused by a falling transition at either input c or d, and then progressing as follows: rising $j \rightarrow$ falling $m \rightarrow$ rising $n \rightarrow$ falling o.

A useful concept is the notion of the *required time*, R, at a node in the circuit. If a circuit is provided with a set of arrival time constraints at each output node, then on the completion of the CPM traversal, one may check whether those constraints are met or not. If they are not met, it is useful to know which parts of the circuit should be sped up and how, and if they are, it may be possible to save design resources by resynthesizing the circuit to slow it down. In either case, the required time at each node in the circuit provides useful information. At any node in the circuit, if the arrival time exceeds the required time, then the path that this node lies on must be sped up.

The computation of the required time proceeds as follows. At each primary output, the rise/fall required times are set according to the specifications provided to the circuit. Next, a backward topological traversal is carried out, processing each gate when the required times at all of its fanouts are known. In essence, this is equivalent to performing a CPM traversal with

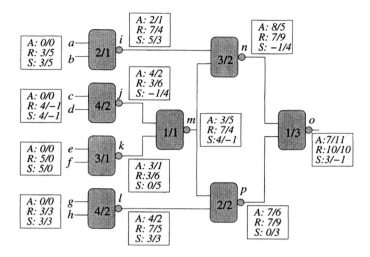

Figure 5.4.   The arrival times, A, the required times, R, and the slacks, S, for the circuit in Figure 5.3. For this example, the values for each input to a gate connected to the primary inputs are identical and are shown only once.

1. the directions of the edges in $G(V, E)$ reversed,

2. the block delays negated, and

3. the required time at the outputs set to the timing specification.

The required rise and fall arrival times for the circuit in Figure 5.3 are shown in Figure 5.4, for a timing specification of 10 units at the primary output. A representative calculation for gate m can be shown as follows: The falling transition at m corresponds to a rising transition at its fanout gates n and p, and the fall required time must satisfy the requirements from both of these gates, and as a result, it is given by the minimum of the values imposed by the two. Each such value is given by the required rise time at the gate output, minus the rise delay of the gate. Therefore, the required time for the falling transition at m is computed as $\min(7 - 3, 7 - 2) = 4$. Similarly, the required time for the rising transition at m is calculated as $\min(9 - 2, 9 - 2) = 7$.

The *slack* associated with each connection is then simply computed as the difference between the required time and the arrival time. A positive slack s at a node implies that the arrival time at that node may be increased by s without affecting the overall delay of the circuit. Such a delay increase will only eat into the slack, and may provide the potential to free up excessive design resources used to build the circuit. The rise and fall slacks at each gate in Figure 5.4, corresponding to a timing constraint of 10 units, are listed next to the gate. For example, the rise delay of gate i in the figure may be slowed down by up to 5 units, and its fall delay by up to 4 units without affecting the circuit delay. This could be accomplished, for example, by replacing the gate by another gate in the library with the same functionality, but higher rise and fall delays. The redistribution of slack plays a vital role in timing analysis, and various methods for this purpose have been presented in, for example, [NBHY89, YS90, SSP02].

It should be noted that the critical path can also be found using the slack information. The lowest slack at the primary output is -1, and by tracing back from the primary output towards the primary inputs and following the path with a slack of -1 yields the critical path of the circuit.

5.2.3   Extensions to more complex cases

For ease of exposition, the example in the previous section contained a number of simplifying assumptions. The critical path method can work under more general problem formulations, and a few of these that are commonly encountered are listed below:

Nonzero arrival times at the primary inputs If the combinational block is a part of a larger circuit, we may have nonzero rise/fall arrival times at the primary inputs. If so, the CPM traversal can be carried out by simply using these values instead of zero as the arrival times at the primary inputs. This is particularly useful in characterizing blocks, and may be used to create hierarchical models.

Minimum delay calculations When the gate delays are fixed, the method described above can easily be adapted to find the minimum, instead of the maximum delay from any input to any output. The only changes in the procedure involve the manner in which an individual block is processed: the earliest arrival time over all inputs is now added to the delay of the block to find the earliest arrival time at its output.

Minmax delay calculations If the gate delay is specified in terms of an interval, $[d_{min}, d_{max}]$, then the minimum and maximum arrival time intervals can be propagated in a similar manner; again, these values may be maintained separately for the rising and falling output transitions. The values of d_{min} and d_{max} can be computed on the fly while processing a gate, as will be explained shortly.

Noninverting gates In the example above, all blocks are assumed to consist of inverting gates, due to which a rising transition at the output is caused by a falling transition at the input and vice versa. In case of a noninverting gate/block, a simple adjustment may be made by realizing that the rise [fall] transition at the output is effected by a rise [fall] transition at the input.

Generalized block delay models If the delay from input pin p to output pin q of a blocks is d_{pq}, and the values of d_{pq} are not all uniform for a block, then the arrival time at the output q can be computed as

$$\max_{\text{all inputs } p} (t_p + d_{pq}),$$

where t_p is the arrival time (of the appropriate polarity) at input p. Note that if $d_{pq} = d \ \forall \ p$, then this simply reduces to the expression,

$$\max_{\text{all inputs } p} (t_p) + d,$$

which was used in the example in Figure 5.3.

Incorporating input signal transition times In the example, the delay of each of the individual gates was prespecified as an input to the problem. However, in practice, the delay of a logic gate depends on factors such as the input transition time (i.e., the 10%-to-90% rise or fall times), which are unknown until the predecessor gate has been processed. This problem is overcome by propagating not only the worst case rise and fall arrival times, but also the input signal transition times corresponding to those arrival times.

Note that this fits in well with the fact that the delay of a gate is only required at the time when it is marked as ready for processing during the CPM traversal. At this time, the input signal arrival times as well as the corresponding signal transition times are available, which implies that the delay of the gate may be computed on the fly, just before it is needed, using techniques such as those presented in Section 4.3 and 4.4.

The last of these is particularly important to factor in, and while the above discussion indicates this may easily be carried out in the context of CPM, such a calculation entails some complexities. Consider for instance two signals arriving at a node, with arrival times of t_1 and t_2, respectively, and transition times of τ_1 and τ_2, respectively. In the most common application of CPM, it is assumed that if $t_1 > t_2$, then the arrival time t_1 is propagated, along with its corresponding transition time, τ_1. However, it is easy to see that if t_1 is *just barely* larger than t_2, but $\tau_1 \ll \tau_2$, then the worst-case transition at the gate output is very likely to be caused by the signal arriving at τ_2. Worst of all, such a scheme that uses t_1 and τ_1 to compute the worst-case delay may very well underestimate the path delay: underestimation is a cardinal sin in timing analysis, where every effort is made to ensure that the estimates are pessimistic (but not unduly so).

A simplistic scheme to overcome this problem may propagate, for each gate, the worst-case arrival time, and the worst-case transition time. However, this may be far too pessimistic, since the worst-case transition time may well be due to an early arriving signal that could never lie on the critical path. This problem was pointed out in [BZS02], which also proposes a method for overcoming it. The essential idea is to selectively propagate a set of arrival times and transition times, using careful pruning to manage the amount of data to be kept track of. Experimental results show that this method provides improved delay estimates with a manageable amount of storage and computation.

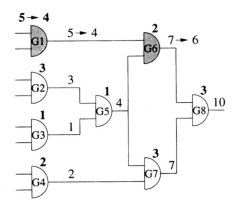

Figure 5.5.   Event-driven static timing analysis.

5.2.4 Incremental timing analysis

During the process of circuit design or optimization, it is frequently the case that a designer or a CAD tool may alter a small part of a circuit and consider the effects of this change on the timing behavior of the circuit. Since the circuit remains largely unchanged, performing a full static timing analysis entails needless computation, since many of the arrival times remain unchanged. The notion of incremental timing analysis provides a way of propagating the changes in circuit timing caused by this alteration only to those parts of the circuit that require it. For example, in the circuit shown in Figure 5.5, if the delay of gate G1 is changed from 5 to 4 units, then its effect is propagated to the next stage of gates, and then on to the next stage, and so on, as long as some arrival time in the circuit is altered. In this case, the arrival times at the shaded gates G1 and G6 are affected, and the propagation stops at gate G8 since its arrival time is unaltered (though in general, such a change may well make its effects felt on the primary outputs). It is easily seen that the incremental analysis here is cheaper than carrying out a fresh new static timing analysis.

This is referred to as *event-driven* propagation: an event is said to occur when the timing information at the input to a gate is changed, and the gate is then processed so that the event is propagated to its outputs. Unlike CPM, in an event-driven approach, it is possible that a gate may be processed more than once. However, when only a small change is made to the circuit, and a small fraction of the gates have their arrival times altered, such a method is highly efficient. In the next section, we will employ an event-driven method for the problem of finding the delays for all primary input-primary output pairs in a combinational circuit, embedding an approach that ensures that each gate is updated only once during the event-driven propagation.

5.2.5   Calculating the worst-case delays from a specific input to all outputs

The CPM procedure described above finds the worst-case delay from any input to any output in a combinational block. A commonly encountered problem is that of finding the maximum delay, $\overline{d}(i,j)$, from *one* specified input i to all outputs j, and a solution was suggested in [Fis90]. This method first set the arrival time at input i to zero, and that at all other inputs to $-\infty$. The resulting signal arrival time at each output j, found using CPM, is the value of $\overline{d}(i,j)$. However, if this procedure were to be performed directly, it would lead to large computation times. This would be accentuated if all input-to-output delays were required, which would mean that this procedure would have to be executed for each input to the combinational block.

For practical circuits, it was observed during the symbolic propagation of constraints in [CSH95] that in most cases, the input to a combinational block exercises only a small fraction of all of the paths between the inputs of the combinational block and the outputs. Based on this observation, an efficient procedure for calculating the values of $\overline{d}(i,j)$ was developed in [Sap96]. It was found that the use of this procedure yielded run-time improvements of several orders of magnitude over the direct multiple applications of CPM described in the previous paragraph.

The topological level, level(k), of each gate k in the circuit is first computed by a single CPM run; this is defined as the largest number of gates from a primary input to the gate, inclusive of the gate, and is found by a topological ordering algorithm. To find $\overline{d}(i,o)$, the largest delay from primary input i to all primary outputs o, we conduct an event-driven CPM-like exercise starting at flip-flop i, as described in the following piece of pseudocode. During the process, we maintain a set of queues, known as level queues. The level queue indexed by k contains all gates at level k that have had an input processed. The pseudocode for the procedure is as shown in Figure 5.6.

At each step, an element from the lowest unprocessed level is plucked from its level queue, and the worst-case delay from input i to the gate output is computed. All of its fanouts are then placed on their corresponding level queues, unless they have already been placed there earlier. Note that by construction, no gate is processed until the delay to all of its inputs that are affected by flip-flop i have been computed, since such inputs must necessarily have a lower level number.

5.3   FALSE PATHS

The true delay of a combinational circuit corresponds to the worst case over all possible logic values that may be applied at the primary inputs. Given that each input can take on one of four values (a steady 0, a steady 1, a $0 \rightarrow 1$ transition and a $1 \rightarrow 0$ transition), the number of possible combinations for a circuit with m inputs is 4^m, which shows an exponential trend. However, it can be verified that the critical path method for finding the delay of a combinational circuit can be completed $O(|V| + |E|)$ time for a timing graph $G = (V, E)$, and in

```
Algorithm ALL_OUTPUT_DELAYS_FROM_SINGLE_INPUT(i)
Find level(k) for each gate k;
Set d̄(i,j) = -∞ for all gates j;
Initialize all level queues to be empty;
currentlevel = 0;
currentgate = input i;
while (currentgate ≠ nil) {
   if (currentgate ≠ i)
      for (all fanins j of currentgate)
         d̄(i,currentgate) = max(d̄(i,j)) + delay_j(currentgate);
   else
         d̄(i,i) = 0;
   for (all fanouts k of currentgate)
      Append gate k to the level queue indexed by level(k),
         if it does not already lie on the queue;
   if (all level queues are not empty) {
      currentlevel = min. level number with nonempty level queue;
      currentgate = head of the level queue for currentlevel;
      delete currentgate from its level queue;
   }
   else
      currentgate = nil;
}
```

Figure 5.6. Pseudocode for an algorithm that efficiently finds all input-output delay pairs for a combinational circuit.

practice, this computation is considerably cheaper on large circuits than a full enumeration. The difference arises because of a specific assumption made by CPM: namely, that the logic function implemented by a gate is inconsequential, and only its delay is relevant. In other words, CPM completely ignores the Boolean relationships in a circuit, and work with purely topological properties. As a result, it is possible that it may not be possible to excite the critical path found by CPM, and in general, the critical path delay found using CPM is pessimistic.

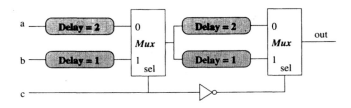

Figure 5.7. An illustration of a circuit with false paths.

As an example, consider the circuit shown in Figure 5.7, with three inputs a, b and c, and one output, out. Assume, for simplicity, that the multiplexer and inverter have zero delays, and that the four blocks whose delays are shown are purely combinational. It can easily be verified that the worst-case delay for this circuit computed using the critical path method, is 4 units. However, by enumerating both possible logic values for the multiplexer, namely, c=0 and c=1, it can be seen the delay in both cases is 3 units, implying that the circuit delay is 3 units. The reason for this discrepancy is simple: the path with a delay of 4 units can never be sensitized because of the restrictions placed by the Boolean dependencies between the inputs.

While many approaches to false path analysis have been proposed, most are rather too complex to be applied in practice. The identification of false paths includes numerous subtleties. Some paths may be statically insensitizable when delays are not considered, but dynamically sensitizable. For instance, if the inverter has a delay of 3 units, then the path of delay 4 units is indeed dynamically sensitizable. Various definitions have been proposed in the literature, and an excellent survey, written at a time when research on false paths was at its most active, in presented in [MB91].

However, by and large, most practical approaches use some designer input and case enumeration, just as in the case of Figure 5.7 where the cases of c=0 and c=1 were enumerated. If a set of user-specified false paths is provided, the techniques listed in Section 5.4 may be applied to identify the longest true path. Alternatively, approaches for pruning user-specified false paths from timing graphs have been presented in [Bel95, GS99, BPD00].

5.4 FINDING THE K MOST CRITICAL PATHS

It is frequently useful to find the k most critical paths in a circuit. For instance, while trying to find the longest sensitizable critical path, one could begin with the longest path and check whether it is a false path or not. If the path is true, the task is done; if not, then the next critical path must be identified, and so on.

A simple algorithm that is easily implemented in the context of the critical path method was presented in [JS91]. This procedure does not require the value of k to be specified *a priori*. It uses the timing graph defined in Section 5.2, associating a delay on each edge that corresponds to the delay from the gate to its fanout, including the gate delay and the interconnect delay[3]. In addition, a source vertex s and a sink vertex t are introduced, as described earlier.

The method proceeds as follows:

- The *maximum delay to a sink* from each node v (*max_delay_to_sink(v)*), defined as the longest delay path from v to the sink node, is determined. This may easily be computed by a backward topological traversal from the primary outputs towards the primary inputs.

- For each connection from a node v to a successor node u_i, if $d(v, u_i)$ is the delay along the edge (v, u_i), the successors are arranged in non-increasing order of the cost function $(d(v, u_i) + max\_delay\_to\_sink(u_i))$,

- The *branch_slack* (abbreviated as *bs* in the figures accompanying this discussion) is computed as the difference between the cost functions of adjacent successors u_i and u_{i+1}, where u_{i+1} is the next node in the sorted list of successors of node v. If the node v has f_v fanouts, then the *branch_slack* for the last node on the list is defined[4] as $d(v, u_{f_v}) + max\_delay\_to\_sink(u_{f_v})$.

As an example, consider the graph shown in Figure 5.8 that shows a part of a circuit. The italicized numbers on each edge indicate the delays associated with each edge, and the bold figures next to each node indicate the *max_delay_to_sink* values. For instance, if the *max_delay_to_sink* for D and G are given to be 10 and 5, respectively, the value of *max_delay_to_sink* for C is computed as max$(10 + 10, 5 + 12) = 20$. The *branch_slack* for the edge (C,D) is therefore 3. The successors of C are sorted in the order of their value of $(d(v, u_i) + max\_delay\_to\_sink(u_i))$, i.e., in the order D, G. Similarly, the *branch_slack* along (B,F) is found to be 1 unit, and the successors of B are sorted in the order C, F.

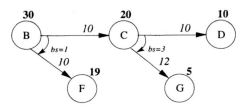

Figure 5.8. An example illustrating the calculation of branch slacks. The number associated with each arc corresponds to its delay, while the boldfaced numbers next to a vertex represent its *max_delay_to_sink* [JS91].

These computations may be used to assist in the enumeration of the k most critical paths. The intuition for the procedure is based on the fact that the delay of the next longest path that branches out from some node v through some edge (v, u) is lower than the longest path through v by the *branch_slack* of (v, u). Each such path must intersect the previous longest path at least two nodes, the source s and the sink t.

The computation proceeds as follows. The longest path in the circuit, $P_1 = (s = v_0, v_1, v_2, \cdots, v_n, v_{n+1} = t)$, is first identified using the techniques described in Section 5.2.1. As this path is identified, the *branch_slack* values at the branch points on this path are used to compute the delays of the corresponding paths, which are all realistic candidates for the next longest path. These delays are stored in an ordered list, and the delay of the next longest path, *next_delay*, is computed using the smallest of these *branch_slacks*. The second-longest path, P_2, is the path with the smallest *branch_slack* on this list. This path branches out from the node v_j on P_1 that has the smallest *branch_slack* for the edge (v_j, u) for some successor node u of v_j. This path is identical to P_1 up to some node v_j, after which the remaining nodes are obtained by identifying the first successor node in the graph until the sink t is reached. Once this path is identified, the *branch_slack* values on the branch points on this path are used to compute the delays for the corresponding candidate path delays, and these are added to the ordered list.

In general, once the j most critical paths have been identified, the $(j + 1)^{\text{th}}$ most critical path is computed by choosing the longest delay path on the ordered list. The branch points off this path are then used to update the ordered list. For efficiency, the ordered list is maintained as a heap [CLR90].

Example: The graph listed in Figure 5.9 shows a DAG on which the k most critical paths are to be found. The first step is to find the values of the *max_delay_to_sink* and *branch_slack* throughout the graph; these are shown in the figure using the same scheme as in Figure 5.8. A trace of the execution that shows the computation of the five most critical paths is shown below.

Critical path #	Path	Delay	Available branch point [*branch_slack*, path delay]	Branch point used (path #)
1	ABFCD	13	A[1,12], B[1,12], C[2,11], F[3,10]	–
2	AEBFCD	12	B[1,11], E[2,10], C[2,10], F[3,9]	A[1,12] (1)
3	ABCD	12	A[1,11], C[2,10]	B[1,12] (1)
4	ABFCGD	11	A[1,10], B[1,10], F[3,8]	C[2,11] (1)
5	AEBCD	11	E[2,9], C[2,9]	B[1,11] (2)

The first path that is identified is the most critical path, ABFCD, with a delay of 13. The available branch points and the delays on the next most critical paths using each of those branch points may be calculated using the corresponding branch slacks as shown, and arranged in the heap in order of the delay. The next critical path is found by taking the largest delay path off the heap. Of the two choices of branch points A and B, A is chosen if it is arranged on the heap before B, leading the the second most critical path. The procedure continues: up to the fourth most critical path, all paths are derived from the most critical path. However, observe that the fifth-most critical path corresponds to the branch slack along B on the second-most critical path.

5.5 SUMMARY

In this chapter, we have developed the material from the earlier chapters to present techniques that find the delay of a combinational block. The critical path method for delay computation was first presented, followed by a discussion on false paths and timing analysis in the presence of false paths.

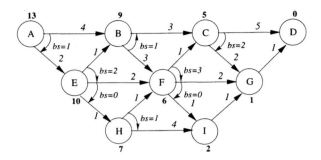

Figure 5.9.   An example on which the k most critical paths are found [JS91].

Notes

1. If some arrival times are negative, the time variable can be shifted to ensure that $a_i \geq 0$ at each primary input.

2. In the context of project management, PERT is a technique that is used to propagate coarse probabilistic delay information through a network, while CPM propagates static delay information. The confusion probably arose because one of the earliest works on timing analysis by Kirkpatrick and Clark [KC66] used PERT for statistical timing analysis (the vintage of this article is apparent from the fact that their illustrative example was that of a ferrite core system design). These principles were extended by later work for constant delay values, without necessarily changing the nomenclature. For an overview of the genesis of timing analysis methods, the reader is referred to an early paper by Hitchcock [Hit82]. As a final note, while statistical timing analysis is newly emerging as a major problem area in its own right, as discussed in Chapter 6, PERT-based approaches may be somewhat simplistic, and unable to model the complexities of the delay variations.

3. An alternative way of building the timing graph is to represent input pins and output pins as vertices. Such a construct permits the specification of unequal pin-to-pin delays in a gate.

4. This definition is necessary since u_{i+1} is undefined for this node.

6 STATISTICAL STATIC TIMING ANALYSIS

6.1 INTRODUCTION

Under true operating conditions, the parameters chosen by the circuit designer are perturbed from their nominal values due to various types of variations. As a consequence, a single SPICE-level transistor or interconnect model (or an abstraction thereof) is seldom an adequate predictor of the exact behavior of a circuit. These sources of variation can broadly be categorized into two classes

Process variations result from perturbations in the fabrication process, due to which the nominal values of parameters such as the effective channel length (L_{eff}), the oxide thickness (t_{ox}), the dopant concentration (N_a), the transistor width (w), the interlayer dielectric (ILD) thickness (t_{ILD}), and the interconnect height and width (h_{int} and w_{int}, respectively).

Environmental variations arise due to changes in the operating environment of the circuit, such as the temperature or variations in the supply voltage (V_{dd} and ground) levels. There is a wide body of work on analysis techniques to determine environmental variations, both for thermal issues [CK00, CS03b, WC02, GS03], and for supply net analysis [SS02].

Both of these types of variations can result in changes in the timing and power characteristics of a circuit.

Process variations can also be classified into the following categories:

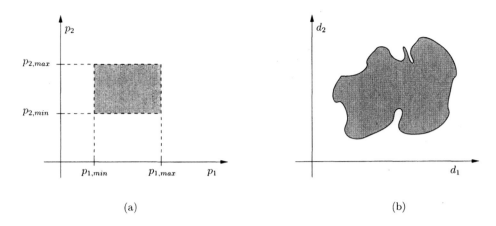

Figure 6.1.   The feasible region in the performance parameter space and (b) the design/process parameter space.

Inter-die variations are the variations from die to die, and affect all the devices on same chip in the same way, e.g., they may cause all of the transistor gate lengths of devices on the same chip to be larger or all of them to be smaller.

Intra-die variations correspond to variability within a single chip, and may affect different devices differently on the same chip, e.g., they may result in some devices having smaller oxide thicknesses than the nominal, while others may have larger oxide thicknesses.

Inter-die variations have been a longstanding design issue, and for several decades, designers have striven to make their circuits robust under the unpredictability of such variations. This has typically been achieved by simulating the design at not just one design point, but at multiple "corners." These corners are chosen to encapsulate the behavior of the circuit under worst-case variations, and have served designers well in the past. In nanometer technologies, designs are increasingly subjected to numerous sources of variation, and these variations are too complex to capture within a small set of process corners.

To illustrate this, consider the design of a typical circuit. The specifications on the circuit are in the form of limits on performance parameters, p_i, such as the delay or the static or dynamic power dissipation, which are dependent on a set of design or process parameters, d_i, such as the transistor width or the oxide thickness. In Figure 6.1(a), we show the behavior of a representative circuit in the performance space of parameters, p_i, whose permissible range of variations lies within a range of $[p_{i,min}, p_{i,max}]$ for each parameter, p_i, which corresponds to a rectangular region. However, in the original space of design parameters, d_i, this may translate into a much more complex geometry, as shown in Figure 6.1(b). This may conservatively be captured in the form of process corners at which the circuit is simulated.

Although it is conceptually possible to model the shape of the feasible region, a large amount of related prior work in the analog context [DH77, AMH91, SVK94] has shown that such approaches can handle problems of only limited dimensionality. This is also borne out by the work in [JKN+03] in the context of digital circuits.

In nanometer technologies, intra-die variations have become significant and can no longer be ignored. As a result, a process corner based methodology, which would simulate the entire chip at a small number of design corners, is no longer sustainable. A true picture of the variations would use one process corner in each region of the chip, but it is clear that the number of simulations would increase exponentially with the number of such regions. This implies that if a small number of process corners are to be chosen, they must be very conservative and pessimistic. For true accuracy, this can be overcome by using a larger number of process corners, but this number may be too large to permit computational efficiency.

Our discussion in this chapter will focus primarily on intra-die variations. Unlike inter-die variations, whose effects can be captured by a small number of static timing analysis (STA) runs at the process corners, a more sophisticated approach is called for in dealing with intra-die variations. This requires an extension of traditional STA techniques to move beyond their deterministic nature. An alternative approach that overcomes these problems is *statistical STA* (SSTA), which treats delays not as fixed numbers, but as probability density functions (PDF's), taking the statistical distribution of parametric variations into consideration while analyzing the circuit.

The sources of these variations may be used to create another taxonomy:

Random variations (as the name implies) depict random behavior that can be characterized in terms of a distribution. This distribution may either be explicit, in terms of a large number of samples provided from fabrication line measurements, or implicit, in terms of a known probability density function (such as a Gaussian or a lognormal distribution) that has been fitted to the measurements. Random variations in some process or environmental parameters (such as those in the temperature, supply voltage, or L_{eff}) can often show some degree of local *spatial correlation*, whereby variations in one transistor in a chip are remarkably similar in nature to those in spatially neighboring transistors, but may differ significantly from those that are far away. Other process parameters (such as t_{ox} and N_a) do not show much spatial correlation at all, so that for all practical purposes, variations in neighboring transistors are uncorrelated.

Systematic variations show predictable variational trends across a chip, and are caused by known physical phenomena during manufacturing. Strictly speaking, environmental changes are entirely predictable, but practically, due to the fact that these may change under a large number (potentially exponential in the number of inputs and internal states) of operating modes

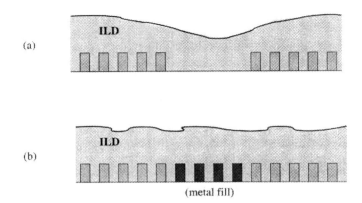

Figure 6.2. An illustration of ILD variations due to metal fill density variations. In (a), the ILD thickness is seen to be reduced in the region with the lower metal density. In (b), dummy metal fills are added so that these deterministic components of the variations are removed, and only the random components of the variations remain.

of a circuit, it is easier to capture them in terms of random variations. Examples of systematic variations include those due to

Spatial intra-chip gate length variability, also known as across-chip linewidt variation (ACLV), which observes systematic changes in the value of L_{eff} across a reticle due to effects such as changes in the stepper-induced illumination and imaging nonuniformity due to lens aberrations [OMC+02].

ILD variations due to the effects of chemical-mechanical polishing (CMP) on metal density patterns: regions that have uniform metal densities tend to have more uniform ILD thicknesses than regions that have nonuniformities. This is illustrated by an example in Figure 6.2.

This chapter will overview the field in the area of SSTA in the presence of intra-die variations. For simplicity, many of these methods assume either a normal distribution for the gate delay, or a discrete probability density function (PDF). The latter have the advantage of being more versatile in describing on-chip variations, but the former can more easily capture spatial correlation structures.

6.2   MODELING PARAMETER VARIATIONS

6.2.1   Components of variations

In general, the intra-chip parametric variation δ can be decomposed into three parts: a deterministic global component, δ_{global}, a deterministic local component δ_{local} and a random component ϵ [LNPS00]:

$$\delta = \delta_{global} + \delta_{local} + \epsilon \qquad (6.1)$$

The global component, δ_{global}, is location-dependent. For example, across the die, it can be modeled by a slanted plane and expressed as a simple function of the die location:

$$\delta_{global}(x, y) \quad = \quad \delta_0 + \delta_x x + \delta_y y \qquad (6.2)$$

where (x, y) is its die location, δ_x and δ_y are the location-dependent gradients of parameter indicating the spatial variations of parameter along the x and y direction respectively.

The local component, δ_{local}, is proximity-dependent and layout-specific. The random residue, ϵ, stands for the random intra-chip variation and is modeled as a random variable with a multivariate normal distribution ϵ to account for the spatial correlation of the intra-chip variation:

$$\epsilon \sim N(0, \Sigma) \qquad (6.3)$$

where Σ is the covariance matrix of the distribution. When the parameter variations are assumed to be uncorrelated, Σ is a diagonal matrix; spatial correlations are captured by the off-diagonal cross-covariance terms in a general Σ matrix using methods such as those described in Section 6.2.2. A fundamental property of covariance matrices says that Σ must be symmetric and positive semidefinite.

If the impact of only the global and random components are considered, then under the model of Equation (6.2), at a given location, the true value of a parameter p at location (x, y) can be modeled as:

$$p \quad = \quad \bar{p} + \delta_x x + \delta_y y + N(0, \Sigma) \qquad (6.4)$$

where $\bar{p}$ is the nominal design parameter value at die location $(0, 0)$.

6.2.2 Spatial correlations

To model the intra-die spatial correlations of parameters, the die region may be partitioned into $nrow \times ncol = n$ grids. Since devices or wires close to each other are more likely to have similar characteristics than those placed far away, it is reasonable to assume perfect correlations among the devices [wires] in the same grid, high correlations among those in close grids and low or zero correlations in far-away grids. For example, in Figure 6.3, gates a and b (whose sizes are shown to be exaggeratedly large) are located in the same grid square, and it is assumed that their parameter variations (such as the variations of their gate length), are always identical. Gates a and c lie in neighboring grids, and their parameter variations are not identical but are highly correlated due to their spatial proximity. For example, when gate a has a larger than nominal gate length, it is highly probable that gate c will have a larger than nominal gate length, and less probable that it will have a smaller than nominal gate length. On the other hand, gates a and d are far away from each other, their parameters are uncorrelated; for example, when gate a has a

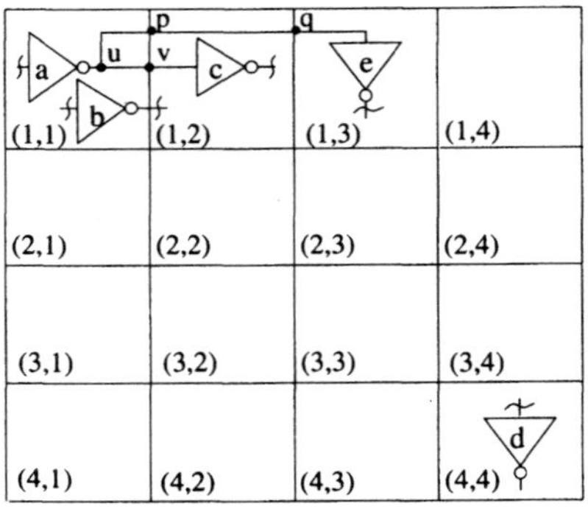

Figure 6.3.   Grid model for spatial correlations [CS03a].

larger than nominal gate length, the gate length for d may be either larger or smaller than nominal.

Under this model, a parameter variation in a single grid at location (x, y) can be modeled using a single random variable $p(x, y)$. For each type of parameter, n random variables are needed, each representing the value of a parameter in one of the n grids.

In addition, it is reasonable to assume that correlation exists only among the same type of parameters in different grids and there is no correlation between different types of parameters. For example, the L_g values for transistors in a grid are correlated with those in nearby grids, but are uncorrelated with other parameters such as T_{ox} or W_{int} in any grid. For each type of parameter, an $n \times n$ correlation matrix, Σ, represents the spatial correlations of such a structure.

An alternative model for spatial correlations was proposed in [ABZ+02, ABZ+03c]. The chip area is divided into several regions using multiple quad-tree partitioning, where at level l, the die area is partitioned into $2^l \times 2^l$ squares; therefore, the uppermost level has just one region, while the lowermost level for a quad-tree of depth k has 4^k regions. A three-level tree is illustrated in Figure 6.4. An independent random variable, $\Delta p_{i,r}$ is associated with each region (i, r) to represent the variations in parameter p in the region at level r. The total variation at the lowest level is then taken to be the sum of the variations of all squares that cover a region.

For example, in Figure 6.4, in region $(2,1)$, if p represents the effective gate length due to intra-die variations, $\Delta L_{eff}(2, 1)$, then

$$\Delta L_{eff}(2, 1) = \Delta L_{0,1} + \Delta L_{1,1} + \Delta L_{2,1} \qquad (6.5)$$

In general, for region (i, j),

$$\Delta p(i, j) = \sum_{0 < l < k, (l,r) \text{ covers } (i,j)} \Delta p_{l,r} \qquad (6.6)$$

It can be shown rather easily that this is a special case of the model of Figure 6.3, and has the advantage of having fewer characterization parameters. On the

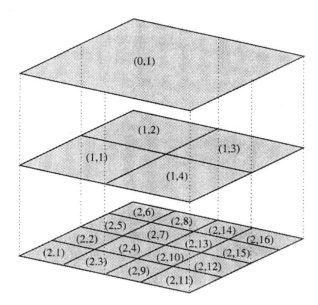

Figure 6.4.   The quadtree model for spatially correlated variations [ABZ$^+$03c].

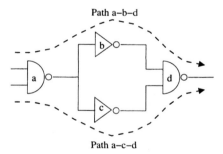

Figure 6.5.   An example to illustrate structural correlations in a circuit.

other hand, it shows marked edge effects that result in smaller correlations between adjacent cells if they fall across the edges of early levels of the quad-tree than those that do not.

6.2.3   Structural correlations

The structure of the circuit can also lead to correlations that must be incorporated in SSTA. Consider the reconvergent fanout structure shown in Figure 6.5. The circuit has two paths, a-b-d and a-c-d. If, for example, we assume that each gate delay is a Gaussian random variable, then the PDF of the delay of each path is easy to compute, since it is the sum of Gaussians, which admits a closed form. However, the circuit delay is the maximum of the delays of these two paths, and these are correlated since the delays of a and d contribute to both paths. It is important to take such structural correlations, which arise due to reconvergences in the circuit, into account while performing SSTA.

6.2.4   Modeling gate/interconnect delay PDF's

We will now show how the variations in the process parameters are translated into PDF's that describe the variations in the gate and interconnect delays that correspond to the weights on the nodes and edges, respectively, of the statistical timing graph.

In Section 6.2.1, the geometrical parameters associated with the gate and interconnect are modeled as normally distributed random variables. Before we introduce how the distributions of gate and interconnect delays will be modeled, let us first consider an arbitrary function $d = f(\mathbf{P})$ that is assumed to be a function on a set of parameters $\mathbf{P}$, where each $p_i \in \mathbf{P}$ is a random variable with a normal distribution given by $p_i \sim N(\mu_{p_i}, \sigma_{p_i})$. We can approximate d linearly using a first order Taylor expansion:

$$d = d_0 + \sum_{\forall \text{ parameters } p_i} \left[\frac{\partial f}{\partial p_i} \right]_0 \Delta p_i \tag{6.7}$$

where d_0 is the nominal value of d, calculated at the nominal values of parameters in the set $\mathbf{P}$, $\left[\frac{\partial f}{\partial p_i} \right]_0$ is computed at the nominal values of p_i, $\Delta p_i = p_i - \mu_{p_i}$ is a normally distributed random variable and $\Delta p_i \sim N(0, \sigma_{p_i})$.

If all of the parameter variations can be modeled by Gaussian distributions, this approximation implies that d is a linear combination of Gaussians, which is therefore Gaussian. Its mean μ_d, and variance σ_d^2 are:

$$\mu_d = d_0 \tag{6.8}$$

$$\sigma_d^2 = \sum_{\forall i} \left[\frac{\partial f}{\partial p_i} \right]_0^2 \sigma_{p_i}^2 + 2 \sum_{\forall i \neq j} cov(p_i, p_j) \tag{6.9}$$

where $cov(p_i, p_j)$ is the covariance of p_i and p_j.

This approximation is valid when Δp_i has relatively small variations, in which domain the first order Taylor expansions is adequate and the approximation is acceptable with little loss of accuracy. This is generally true of the impact of intra-chip variations on delay, where the process parameter variations are relatively small in comparison with the nominal values, and the function changes by a small amount under this perturbation. For this reason, the gate and interconnect delays, as functions of the process parameters, can be approximated as a normal distributions when the parameter variations are assumed to be normal.

Computing the PDF of gate delay   For a multiple-input gate, the pin-to-pin delay of the gate differs at different input pins. Let $d_{gate}^{pin_i}$ be the delay of the gate from the ith input to the output. In general, $d_{gate}^{pin_i}$ can be written as a function of the process parameters $\mathbf{P}$ of the gate, the loading capacitance of the driving interconnect tree $\mathbf{C}_w$ and the succeeding gates that it drives $\mathbf{C}_g$, and the input signal transition time S_{in} at this input pin of the gate:

$$d_{gate}^{pin_i} = d(\mathbf{P}, \mathbf{C}_w, \mathbf{C}_g, S_{in}) \qquad (6.10)$$

The sensitivities of the gate delay to the process parameters can be found applying the chain rule for computing derivatives.

Since the gate delay $d_{gate}^{pin_i}$ differs at the different input pins, in conventional static timing analysis, S_{out} is set to $d_{gate}^{pin_i}$ if the path ending at the output of the gate traversing the ith input pin has the longest path delay. In statistical static timing analysis, each of the paths through different gate input pins has a certain probability to be the longest path. Therefore, S_{out} should be computed as a weighted sum of the distributions of the gate delays $d_{gate}^{pin_i}$, where the weight equals the probability that the path through the ith pin is the longest among all others:

$$S_{out} = \sum_{\forall \text{ input pins } i} \{ Prob[d_{path_i} > \max_{\forall j \neq i}(d_{path_j})] \times d_{gate}^{pin_i} \} \qquad (6.11)$$

where d_{path_i} is the distribution of path delay at the gate output through the ith input pin. The calculation of d_{path_i} and $\max_{\forall j \neq i}(d_{path_j})$ can be achieved by the "sum" and the "max" operators, as discussed in Section 5.2.1. It is clearly to see that S_{out} is now approximated as a normal distribution, since it is as a weighted sum of normal distributions $d_{gate}^{pin_i}$. Using the formulation above, the derivatives of S_{out} to the process parameters can also be computed through the weighted sum of the derivatives of $d_{gate}^{pin_i}$ to the process parameters.

6.3 EARLY WORK ON STATISTICAL STA

Early work by [DIY91] presented a method for performing statistical timing analysis while including structural Boolean properties of a combinational circuit. The approach used a discrete PDF and encoded both the delay and logic behavior of the circuit into a Boolean expression that was subsequently simplified using a BDD representation. Although the results of this method were shown on small circuits, a notable observation was related to the computation of the signal probability (i.e., the probability that the signal is at logic 1) at the output of a gate.

Notationally, let $P_a(t)$ be the probability that a line a is at logic 1 at time t. Temporarily assuming that each gate has zero delay, the line probability at the gate output can be calculated for various gates. For example, if the inputs of the gate are $x_1, \cdots, x_k$ and the output is y, then

$$
\begin{aligned}
\text{Inverter:} \quad & P_{y0}(t) = 1 - P_{x_1}(t) \\
k\text{-input Nand:} \quad & P_{y0}(t) = P_{x_1}(t) \times P_{x_2}(t) \times \cdots \times P_{x_k}(t) \\
k\text{-input Nor:} \quad & P_{y0}(t) = 1 - P_{x_1}(t) \times P_{x_2}(t) \times \cdots \times P_{x_k}(t),
\end{aligned}
$$

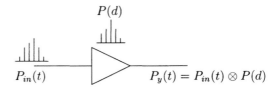

Figure 6.6.   For a one-input gate, the PDF of the output arrival time can be calculated by convolving the PDF of the input arrival time, $P_{in}(t)$, and the PDF of the gate delay, $P(d)$ [DIY91].

which is correct if the circuit has no reconvergences. Now, instead of a zero delay, if a gate has a discrete delay PDF that dictates that its probability of having a delay of d is $P(d)$, and if the signal probability on line i at time t is $P_i(t)$, then the PDF of the line probability at the output can be found using a discrete convolution as

$$P_y(t) = \sum_{d=d_{min}}^{d_{max}} P(d) \times P_{y0}(t - d) \qquad (6.12)$$

This is illustrated in Figure 6.6. The convolution result has since been independently rediscovered and applied by many researchers in the area of SSTA. For continuous PDFs, the same reasoning may be applied to render the equation above in the form of a continuous convolution.

The work in [JMDK93] presented a symbolic simulation method for statistical timing analysis. A particularly notable contribution that has since been used in other work is the idea of using interval analysis to generate pruning strategies to remove paths that can never be (or have a very low likelihood of being) critical. Unlike [DIY91], this method demonstrated results on large benchmark circuits, albeit under interval-based delay models.

A broad taxonomy of SSTA techniques can be made on the basis of the following criteria:

Path-based vs. block-based methods Path-based methods attempt to find the probability distribution of the delay on a path-by-path basis, and eventually performing a "max" operation to find the delay distribution of the circuit. If the number of paths to be considered is small, path-based methods can be effective. In practice, however, the number of paths to be considered in many circuits is exponential in the number of gates. On the other hand, block-based methods perform a topological CPM-like traversal, processing each gate once, and are potentially much faster.

Continuous vs. discrete PDFs The variations in the gates may be modeled using continuous or discrete PDFs. The former have the useful property of providing a compact closed form, while the latter may be more data-intensive, particularly as the number of terms can increase exponentially after repeated convolution operations. However, discrete methods have the advantage of being more general, and the number of data points can, in practice, be limited to a manageable number through clever heuristics.

The Monte Carlo method is probably the simplest method, in terms of implementation, for performing statistical timing analysis. Given an arbitrary delay distribution, the method generates sample points and runs a static timing analyzer at each such point, and aggregates the results to find the delay distribution. The advantages of this method lie in its ease of implementation and its generality in being able to handle the complexities of variations. For example, spatial correlations are easily incorporated, since all that is required is the generation of a sample point on a correlated distribution. Such a method is very compatible with the data brought in from the fab line, which are essentially in the form of sample points for the simulation. Its major disadvantage is, of course, the extremely large run-times that it requires.

6.4 STATISTICAL STA IN THE ABSENCE OF SPATIAL CORRELATION

6.4.1 Continuous PDFs

While the methods outlined in the previous section focused on solving the statistical timing problem together with false path detection, the first method for statistical static timing analysis to successfully process large benchmarks under probabilistic delay models was proposed by Berkelaar in [Ber97]. In the spirit of static timing analysis, this approach was purely topological, and ignored the Boolean structure of the circuit. The method assumed that each gate in the circuit has a delay distribution that is described by a Gaussian PDF.

The essential operations in STA can be distilled into two types:

- A gate is being processed in STA when the arrival times of all inputs are known, at which time the candidate delay values at the output are computed using the "sum" operation that adds the delay at each input with the input-to-output pin delay.

- Once these candidate delays have been found, the "max" operation is applied to determine the maximum arrival time at the output.

In SSTA, the operations are identical to STA; the difference is that the pin-to-pin delays and the arrival times are PDFs instead of single numbers.

Berkelaar's approach maintains an invariant that expresses all arrival times as Gaussians. As a consequence, since the gate delays are Gaussian, the "sum" operation is merely an addition of Gaussians, which is well known to be a Gaussian. The computation of the max function, however, poses greater problems. The set of candidate delays are all Gaussian, so that this function must find the maximum of Gaussians. In general, the maximum of two Gaussians is *not* a Gaussian. However, given the intuition that if a and b are Gaussian random variables, if $a \gg b$, $\max(a, b) = a$, a Gaussian; if $a = b$, $\max(a, b) = a = b$, a Guassian, it may be reasonable to approximate this maximum using a Gaussian. This method was applied and found to be highly effective in [Ber97]. It was suggested there that a statistical sampling approach could be used to approximate the mean and variance of the distribution; alternatively, this information could be embedded in look-up tables. In later work in [JB00], a precise closed-form approximation for the mean and variance, based on [Pap91], was utilized.

6.4.2 Discrete PDFs

Statistical timing analysis has also been pursued in the test community. The approach in [CW00] takes into account capacitive coupling and intra-die process variation to estimate the worst case delay of critical path. The technique in [LCKK01] uses a discrete PDF based method to propagate discrete PDFs through the circuit, developing effective heuristics to overcome inaccuracies in event propagation due to reconvergences. The work in [LKWC02] uses a Monte Carlo sampling-based framework to analyze circuit timing on a set of selected sensitizable true paths. Similar lines of investigation as [LCKK01, LKWC02] have also been independently pursued in [Nai02].

The work in [DK03] develops an efficient algebra for the computation of discrete probabilities in a block-based approach. For a single input gate, the convolution described in Equation (6.12) may be used to propagate a discrete PDF forward, except that instead of propagating the PDF of the signal probability forward, we now propagate the PDF of the arrival time instead. For multiinput gates, a "max" operation must be carried out: for a k-input gate with arrival times $m_i, i = 1, \cdots, k$, and input-to-output delays $d_i, i = 1, \cdots, k$, the arrival time at the output is found as the PDF of

$$m_{out} = \max_{i=1 \text{ to } k} (m_i + d_i) \qquad (6.13)$$

Instead of a discrete PDF, this method uses a piece-wise constant PDF, which translates to a piecewise-linear cumulative density function (CDF); recall that the CDF is simply the integral of the PDF.

The CDF of m_{out} is easily computed given the PDFs of the m_is and d_is. The PDF of $m_i + d_i$ can be obtained by convolving their respective PDFs, i.e.,

$$\text{PDF}(m_i + d_i) = \text{PDF}(m_i) \otimes \text{PDF}(d_i) \qquad (6.14)$$

where $\otimes$ represents the convolution operator. It can be shown that the CDF of this sum is given by

$$\text{CDF}(m_i + d_i) = \text{CDF}(m_i) \otimes \text{PDF}(d_i) = \text{PDF}(m_i) \otimes \text{CDF}(d_i) \qquad (6.15)$$

The CDF of the maximum of a set of *independent* random variables is easily verified to simply be the product of the CDFs, so that we obtain

$$\text{CDF}\left(\max_{i=1 \text{ to } k} (m_i + d_i)\right) = \prod_{i=1 \text{ to } k} (\text{CDF}(m_i) \otimes \text{PDF}(d_i)) \qquad (6.16)$$

The term in the inner parentheses is the product of a piecewise linear CDF with a piecewise constant PDF, which is piecewise linear, and therefore the CDF is found by multiplying a set of piecewise linear terms, which yields a piecewise quadratic. The resulting quadratic is approximated by a piecewise linear CDF, and the process continues as blocks are processed in CPM-like order. The technique also has some mechanisms for considering the effects of structural correlations, and the reader is referred to [DK03] for a detailed discussion.

6.4.3 Bounding methods

The work in [ABZV03a, ABZV03b, AZB03] uses bounding techniques to arrive at the delay distribution of a circuit. The method is applicable to either continuous or discrete PDFs, and at its core, it is based on reduction of the circuit to easily computable forms.

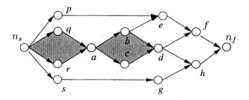

Figure 6.7.   Identifying structural dependencies in the circuit graph [AZB03].

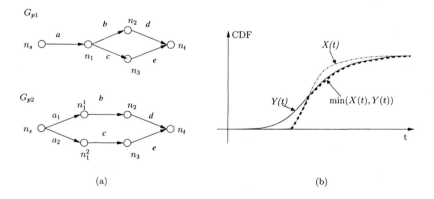

Figure 6.8.   Illustration of the computation of (a) upper and (b) lower bounds for the CDF $X(t)$ for paths $a - b - d$ and $a - c - e$ [AZB03].

The timing graph may be represented by a set of vertices that correspond to gate inputs and outputs and primary inputs/outputs, with edges between the interconnected gates and gate input-to-output pin connections. For such graphs, the PDFs for series-connected edges with a single fanin can be processed by a convolution, and for parallel-connected edges, the CDF may be computed by taking the product of the CDFs of the incoming edges (assuming that these edges are statistically independent). For reconvergent subgraphs that are in neither of these forms, in principle, the computation can be carried out by a path enumeration over the subgraph. However, this is liable to be computationally expensive and is not a feasible option, particularly for reconvergences with a large depth.

Therefore, the method identifies dependence sets corresponding to such reconvergences and employs bounding techniques to approximate the PDFs. The informal definition of a dependence set is illustrated in Figure 6.7, where n_s is a dummy source node and n_t is a dummy sink node. For nodes f and h, the nodes in the shaded region correspond to the transitive fanin that the two nodes share. The set of nodes within this structure that fan out to the remainder of the transitive fanin of f is the set $\{b, d\}$, and this constitutes the dependence set of f. Similarly, the dependence set of h is $\{d\}$. The method proceeds by propagating the arrival time PDFs to the first dependence node, and using conditional probabilities to perform the enumeration. More detailed enumeration schemes that enumerate over different parts of a CDF are also provided.

Since enumeration can be a costly procedure, a bounding technique is introduced. The principles behind bounding are simple to understand. Consider the graph G_{P1} shown in Figure 6.8(a). It is proven in [AZB03] that an *upper bound* on the CDF of the graph G_{P1} in the figure is provided by the CDF of the graph G_{P2}, which simply splits the node n_1 and carries out the PDF propagation by ignoring the structural correlation between the $a - b - d$ path and the $a - c - e$ path. A *lower bound* on the CDFs for two dependent arrival times can be found by simply using the envelope of their CDFs using the min operator on the original graph (without node splitting), and this is shown in Figure 6.8(b).

The overall procedure uses a heuristic method for merit computation for a dependence node, to determine whether it should be enumerated or bounded. Experimental results on benchmark circuits show that these bounds are very successful in generating tight estimates of the CDF.

6.5   STATISTICAL STA UNDER SPATIAL CORRELATIONS

In cases where significant spatial correlations exist, it is important to take these into account. Figure 6.5 shows a comparison of the PDF yielded by an SSTA technique that is unaware of spatial correlations, as compared with a Monte Carlo simulation that incorporates these spatial correlations, and clearly shows a large difference. This motivates the need for developing methods that can handle these dependencies.

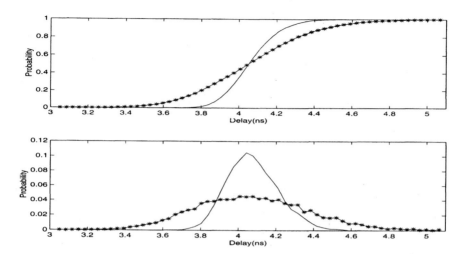

Figure 6.9. A comparison of the results of SSTA when the random variables are spatially correlated. The line on which points are marked with stars represents the accurate results obtained by a lengthy Monte Carlo simulation, and the the solid curve shows the results when spatial correlations are entirely ignored. The upper plot shows the CDFs, and the lower plot, the PDFs [CS03a].

6.5.1 An overview of approaches

The approach described in [Ber97] was extended to handle intra-gate spatial correlations in [TTF00]. This work assumed that the parameters associated with transistors within a gate are correlated, but transistors in different gates have uncorrelated parameters. A notable feature of this work was the use of an approximation technique from [Cla61] (detailed in Appendix B) that provides a closed-form formula to approximate the maximum of two correlated Gaussian random variables as a Gaussian. More recently, the intra-gate variability problem has also been addressed in [OYO03] using a linear response surface-based technique to generate a gate delay model for an individual cell, including sensitivity information.

The work by Orshansky and Keutzer in [OK02] suggested a method for finding the max of path delay PDFs. The covariances between paths are first calculated based on the delay covariances for pairs of gates on the paths, and an analytic method is then used to derive lower and upper bounds of circuit delay. However, the results were demonstrated on small structures and the pairwise computation of covariances could be a potential problem when a large number of paths is to be considered.

Agarwal *et al.* [ABZ+02, ABZ+03c] proposed the quadtree-based spatial correlation model discussed in Section 6.2.2, and show how the delay of a single path can easily be calculated under this model. The essential idea is that since the change in the effective length for a gate in region (i, j) is given by

$$\Delta L(i, j) = \sum_{0<l<k,(l,r) \text{ covers } (i,j)} \Delta L(l, r), \qquad (6.17)$$

the delay can be written in the form

$$\Delta D_{\text{path}} = \sum_{g \in \text{ path}} K_g \sum_{0<l<k,(p,q) \text{ intersects } g} \Delta L(p, q), \qquad (6.18)$$

where the K_g variables correspond to the sensitivity of the delay to the L variables. In other words, the change in delay can be written as a weighted sum of the ΔL variables.

However, the PDF for a critical path may not be a good predictor of the distribution of the circuit delay (which is the maximum of all path delays). Later work by the same authors in [ABZ03b] derived an upper bound on the CDF for the spatially correlated case, based on this quadtree model. Simplifying the notation to represent the length variables as ΔL_i, each arrival time in a CPM-like traversal may be written as

$$D_{\text{path } j} = D_{nom,j} + \sum_i \alpha_{i,j} \Delta L_i + \Delta D_{\text{random},j} \qquad (6.19)$$

where $\alpha_{i,j}$ is the sensitivity of the delay of path j to parameter L_i. To find the max of two such arrival times, the following inequality is used:

$$\max\left(\sum_{i=1}^n a_i, \sum_{i=1}^n x_i\right) \leq \sum_{i=1}^n \max(a_i, x_i) \qquad (6.20)$$

A heuristic technique is used to propagate arrival times during the CPM traversal, using the bound to merge groups of arrival times when possible, and thus propagating a set of arrival times. Experimental results in [ABZ03b] show some success for this approach, but the accuracy near the tail of the CDF is inadequate. It must be stressed, however, that research is still ongoing, and the reader is invited to look out for more recent results that these authors will have published subsequent to [ABZ03b].

A final method for SSTA under spatial correlations is due to [CS03a], and we will discuss this method in detail in the next section.

6.5.2   A principal component analysis based approach

The approach in [CS03a] is based on the application of principal component analysis (PCA) techniques [Mor76] to convert a set of correlated random variables into a set of uncorrelated variables in a transformed space; the PCA step can be performed as a preprocessing step for a design. The complexity of this method is n times the complexity of CPM, where n is the number of squares in the grid, plus the complexity of finding the principal components, which requires very low runtimes in practice. The overall CPU times for this method have been shown to be low, and the method yields high accuracy results.

Given a set of correlated random variables $\mathbf{X}$ with a covariance matrix Σ, the PCA method transforms the set $\mathbf{X}$ into a set of mutually orthogonal random variables, $\mathbf{X}'$, such that each member of $\mathbf{X}'$ has zero mean and unit variance. The elements of the set $\mathbf{X}'$ are called principal components in PCA, and the size of $\mathbf{X}'$ is no larger than the size of $\mathbf{X}$. Any variable $x_i \in \mathbf{X}$ can then be expressed in terms of the principal components $\mathbf{X}'$ as follows:

$$x_i = (\sum_j \sqrt{\lambda_i} \cdot v_{ij} \cdot x_j')\sigma_i + \mu_i \qquad (6.21)$$

where x_j' is a principal component in set $\mathbf{X}'$, λ_i is the ith eigenvalue of the covariance matrix Σ, v_{ij} is the ith element of the jth eigenvector of Σ, and σ_i and μ_i are, respectively, the mean and standard deviation of x_i.

For instance, let $\mathbf{L}_g$ be the set of random variables representing transistor gate length variations in all grids and the set of random variables is of multivariate normal distribution with covariance matrix Σ_{L_g}. Let $\mathbf{L}'_g$ be the set of principal components computed by PCA. Then any $L_g^i \in \mathbf{L}_g$ representing the variation of transistor gate length of the ith grid can then be expressed as a linear function of the principal components:

$$L_g^i = \mu_{L_g^i} + a_{i1} \times l_g'^1 + \cdots + a_{it} \times l_g'^t \qquad (6.22)$$

where $\mu_{L_g^i}$ is the mean of L_g^i, $l_g'^i$ is a principal component in $\mathbf{L}'_g$, all $l_g'^i$ are independent with zero means and unit variances, and t is the total number of principal components in $\mathbf{L}'_g$.

Superposing the set of rotated random variables of parameters on the random variables in gate or interconnect delay as in Equation (6.7), the delay may then be written as a linear combination of the principal components

$$d = d_0 + k_1 \times p_1' + \cdots + k_m \times p_m' \qquad (6.23)$$

where $p_i' \in \mathbf{P}'$ and $\mathbf{P}' = \mathbf{L}'_g \cup \mathbf{W}'_g \cup \mathbf{T}'_{ox} \cup \mathbf{N}'_a \cup \mathbf{W}'_{int_l} \cup \mathbf{T}'_{int_l} \cup \mathbf{H}'_{ILD_l}$ and m is the size of $\mathbf{P}'$.

Since all of the principal components p_i' that appear in Equation (6.23) are independent, the following properties ensue:

- The variance of d is given by

$$\sigma_d^2 = \sum_{i=1}^{m} k_i^2 \qquad (6.24)$$

- The covariance between d and any principal component p_i' is given by:

$$cov(d, p_i') = k_i \sigma_{p_i'}^2 = k_i \qquad (6.25)$$

- For two random variables, d_i and d_j, given by

$$d_i = d_i^0 + k_{i1} \times p_1' + \cdots + k_{im} \times p_m' \qquad (6.26)$$
$$d_j = d_j^0 + k_{j1} \times p_1' + \cdots + k_{jm} \times p_m' \qquad (6.27)$$

The covariance of d_i and d_j, $cov(d_i, d_j)$, can be computed as

$$cov(d_i, d_j) = \sum_{r=1}^{m} k_{ir} k_{jr} \qquad (6.28)$$

In other words, the number of multiplications is linear in the dimension of the space, since orthogonality of the principal components implies that the products of terms k_{ir} and k_{js} for $r \neq s$ need not be considered.

The work in [CS03a] uses these properties to perform SSTA under the general spatial correlation model of Figure 6.3. The method assumes that the fundamental process parameters are in the form of correlated Gaussians, so that the delay, given by Equation (6.7) is a weighted sum of Gaussians, which is therefore Gaussian.

As in the work of Berkelaar, this method maintains the invariant that all arrival times are approximated as Gaussians, although in this case the Gaussians are correlated and are represented in terms of their principal components. Since the delays are considered as correlated Gaussians, the sum and max operations that underlie this block-based CPM-like traversal must yield Gaussians in the form of principal components.

The computation of the distribution of the sum function, $d_{sum} = \sum_{i=1}^{n} d_i$, is simple. Since this function is a linear combination of normally distributed random variables, d_{sum} is a normal distribution whose mean, μ_{dsum}, and variance, σ_{dsum}^2, are given by

$$\mu_{d_{sum}} = \sum_{i=1}^{n} d_i^0 \qquad (6.29)$$
$$\sigma_{d_{sum}}^2 = \sum_{j=1}^{m} \sum_{i=1}^{n} k_{ij}^2 \qquad (6.30)$$

Strictly speaking, the max function of n normally distributed random variables, $d_{max} = max(d_1, \cdots, d_n)$, is not Gaussian; however, like [Ber97], it is approximated as one. The approximation here is in the form of a correlated Gaussian, and the procedure in Appendix B is employed. The result is characterized in terms of its principal components, so that it is enough to find the mean of the max function and the coeficients associated with the principal components.

The utility of using principal components is twofold:

- As described earlier, it implies that covariance calculations between paths are of linear complexity in the number of variables, obviating the need for the expensive pair-wise delay computation methods used in other methods.

Input: Process parameter variations
Output: Distribution of circuit delay

1. Partition the chip into $n = nrow \times ncol$ grids, each modeled by spatially correlated variables.

2. For each type of parameter, determine the n jointly normally distributed random variables and the corresponding covariance matrix.

3. Perform an orthogonal transformation to represent each random variable with a set of principal components.

4. For each gate and net connection, model their delays as linear combinations of the principal components generated in step 3.

5. Using "sum" and "max" functions on Gaussian random variables, perform a CPM-like traversal on the graph to find the distribution of the statistical longest path. This distribution achieved is the circuit delay distribution.

Figure 6.10.   Overall flow of the PCA-based statistical timing analysis method.

- Structural correlations are automatically accounted for, since all the requisite information required to model these correlations is embedded in the principal components.

The overall flow of the algorithm is shown in Figure 6.10. To further speed up the process, several techniques may be used:

1. Before running the statistical timing analyzer, one run of deterministic STA is performed to determine loose bounds on the best-case and worst-case delays for all paths. Any path whose worst-case delay is less than the best-case delay of the longest path will never be critical, and edges that lie only on such paths can safely be removed.

2. During the "max" operation of statistical STA, if the value of $mean + 3 \cdot \sigma$ of one path has a lower delay than the value of $mean - 3 \cdot \sigma$ of another path, the max function can be calculated by ignoring the path with lower delay.

Although the above exposition has focused on handling spatially correlated variables, it is equally easy to incorporate uncorrelated terms in this framework. Only spatially correlated variables are decomposed into principal components, and any uncorrelated variables remain as they are. Therefore, in a case where no variables are spatially correlated, the approach reduces to Berkelaar's method in [Ber97]. However, heuristics from the other approaches in Section 6.4 may be used to improve the modeling of structural correlations.

6.6 SUMMARY

Statistical timing analysis is an area of growing importance in nanometer technologies, as the uncertainties associated with process and environmental variations increase, and this chapter has captured some of the major efforts in this area. This remains a very active field of research, and there is likely to be a great deal of new research to be found in conferences and journals after this book is published.

In addition to the statistical analysis of combinational circuits, a good deal of work has been carried out in analyzing the effect of variations on clock skew. Although we will not treat this subject in this book, the reader is referred to [LNPS00, HN01, JH01, ABZ03a] for details.

7 TIMING ANALYSIS FOR SEQUENTIAL CIRCUITS

7.1 INTRODUCTION

A general sequential circuit is a network of computational nodes (gates) and memory elements (registers). The computational nodes may be conceptualized as being clustered together in an acyclic network of gates that forms a combinational logic circuit. A cyclic path in the direction of signal propagation is permitted in the sequential circuit only if it contains at least one register[1]. In general, it is possible to represent any sequential circuit in terms of the schematic shown in Figure 7.1, which has I inputs, O outputs and M registers. The registers outputs feed into the combinational logic which, in turn, feeds the register inputs. Thus, the combinational logic has $I + M$ inputs and $O + M$ outputs.

Pipelined systems , such as the one shown in Figure 7.2, are a special subset of the class of general sequential circuits, and are commonly used in datapaths. A pipelined system uses registers to capture the output of each logic stage at the end of each clock period. Data proceeds through each combinational block, or pipeline stage, until it reaches the end of the pipeline, with the registers serving to isolate individual stages so that they may parallelly process data corresponding to different data sets, which increases the data throughput rate of the system. While many pipelines have no feedback, the definition of a pipeline does not preclude the use of feedback, and algorithms for scheduling data into a pipeline with feedback have been devised. Acyclic pipelines form an

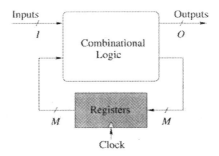

Figure 7.1.   A general sequential circuit.

important subclass of circuits and are popularly used in data-oriented designs as they are much simpler to design, analyze and optimize than general sequential circuits.

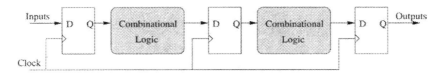

Figure 7.2.   An schematic of a pipelined system. If the combinational logic blocks have multiple inputs, each flip-flop may be replaced by a bank of flip-flops, one for each input, that separates the pipeline stages.

The functionality of registers plays a key role in determining the behavior of any sequential circuit, and there are several choices of register type that are available to a designer. The behavior of each register is controlled by the clock signal, and depending on the state of the clock, the data at the register input is either isolated from its output, or transmitted to the output. The types of registers that may be used in a synchronous system are differentiated by the manner in which they use the clock to transmit data from the input to the output.

Level-clocked latches These are commonly referred to merely as latches, and permit data to be transmitted from the input to the output whenever the clock is high. During this time, the latch is said to be "transparent."

Edge-triggered flip-flops These are commonly called flip-flops (FF's), and use the clock edge to determine when the data is transmitted to the output. In a positive [negative] edge-triggered FF, data is transmitted from the input to the output when the clock makes a transition from 0 to 1 [1 to 0]. FF's can typically be constructed by connecting two level-clocked latches in a master-slave fashion; for details, see [WE93].

A circuit in which all memory elements are level-clocked latches, is commonly referred to as a level-clocked circuit, and a circuit composed of edge-triggered FF's is called an edge-triggered circuit.

In our discussions, we will primarily deal with edge-triggered D flip-flops and level-clocked D latches as representative memory elements. Other types of storage elements that may be used include the RS latch, the JK flip-flop, and the T flip-flop, but we will not address these specifically in our discussion. A description of the transistor-level details of these registers is beyond the scope of this book, but we point out that various choices are available, and they influence the number of transistors used, the number of clock signals to be routed, and the area of and power dissipated by the chip [WE93].

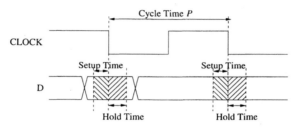

Figure 7.3.   Illustration of the clocking parameters for a register.

7.2 CLOCKING DISCIPLINES: EDGE-TRIGGERED CIRCUITS

We will now overview the timing requirements for edge-triggered sequential circuits, which consist of combinational blocks that lie between D flip-flops. The basic parameters associated with a flip-flop can be summarized as follows:

- The data input of the register, commonly referred to as the D input, must receive incoming data at a time that is at least T_{setup} units before the onset of the latching edge of the clock. The data will then be available at the output node, Q, after the latching edge. The quantity, T_{setup}, is referred to as the setup time of the flip-flop.

- The input, D, must be kept stable for a time of T_{hold} units, where T_{hold} is called the hold time, so that the data is allowed to be stored correctly in the flip-flop.

- Each latch has a delay between the time the data and clock are both available at the input, and the time when it is latched; this is referred to as the clock-to-Q delay, T_q.

In the edge-triggered scenario, let us consider two FF's, i and j, connected only by purely combinational paths. Over all such paths $i \rightsquigarrow j$, let the largest delay from FF i to FF j be $\overline{d}(i,j)$, and the smallest delay be $\underline{d}(i,j)$. Therefore, for any path $i \rightsquigarrow j$ with delay $d(i,j)$, it must be true that

$$\underline{d}(i,j) \leq d(i,j) \leq \overline{d}(i,j)$$

We will denote the setup time, hold time, and the maximum and minimum clock-to-Q delay of any arbitrary FF k as T_{s_k}, T_{h_k}, and Δ_k and δ_k, respectively. For a negative edge-triggered register, the setup and hold time requirements are illustrated in Figure 7.3. The clock is a periodic waveform that repeats after every P units of time, called the clock period or the cycle time.

The data is available at the launching FF, i, after the clock-to-q delay, and will arrive at the latching FF, j, at a time no later than $\Delta_i + \overline{d}(i,j)$. For correct clocking, the data is required arrive one setup time before the latching edge of the clock at FF j as shown in Figure 7.3, i.e, at a time no later than $P - T_{s_j}$. This leads to the following constraint:

$$\Delta_i + \overline{d}(i,j) \leq P - T_{s_j}$$
$$\text{i.e., } \overline{d}(i,j) \leq P - T_{s_j} - \Delta_i \tag{7.1}$$

For obvious reasons, this constraint is often referred to as the *setup time constraint*. Since this requirement places an upper bound on the delay of a combinational path, it is also called the *long path constraint*. A third name attributable to this is the *zero clocking constraint*, because the data will not arrive in time to be latched at the next clock period if the combinational delay does not satisfy this constraint.

The data must be stable for an interval that is at least as long as the hold time after the clock edge, if it is to be correctly captured by the FF. Hence, it is essential to ensure that the new data does not arrive at FF j before time T_{h_j}. Since the earliest time that the incoming data can arrive is $\delta_i + \underline{d}(i,j)$, this gives us the following *hold time constraint*:

$$\delta_i + \underline{d}(i,j) \geq T_{h_j}$$
$$\text{i.e., } \underline{d}(i,j) \geq T_{h_j} - \delta_i \tag{7.2}$$

Since this constraint puts a lower bound on the combinational delay on a path, it is referred to as a *short path constraint*. If this constraint is violated, then the data in the current clock cycle is corrupted by the data from the next clock cycle; as a result, data is latched twice instead of once in a clock cycle, and hence it is also called the *double clocking constraint*. Notice that if the minimum clock-to-Q delay of FF i is greater than the hold time of FF j, i.e., $\delta_i \geq T_{h_j}$ (this condition is not always true in practice), then the right hand side of the constraint is negative. In this case, since $\underline{d}(i,j) \geq 0$, the short path constraint is always satisfied.

An important observation is the both the long path and the short path constraints refer to combinational paths that lie between flip-flops. Therefore, for timing verification of edge-triggered circuits, it is possible to decompose the circuit into combinational blocks, and to verify the validity of the constraints on each such block independently. As we will see shortly, this is not the case for level-clocked circuits, which present a greater complexity to the timing verifier.

7.3   CLOCKING DISCIPLINES: LEVEL-CLOCKED CIRCUITS

7.3.1   Introduction

As mentioned in Section 7.1, unlike an edge-triggered FF, a level-clocked latch is transparent during the active period of the clock. This makes the analysis and design of level-clocked circuits more complex than edge-triggered circuits, since combinational blocks are not insulated from each other by the memory elements, and multicycle paths are possible when the data is latched during the transparent phase of the clock. Even though this transparent nature introduces an additional level of complexity to the design phase, level-clocked circuits are often used for high-performance designs since they offer more flexibility than edge-triggered circuits, both in terms of the minimum clock period achievable and the minimum number of memory elements required.

As an illustration of this notion, consider the simple circuit in Figure 7.4 with unit delay gates and a single-phase clocking scheme with a 50% duty cycle. Let us assume that the data signals are available at the primary inputs at the falling edge of the clock, and must arrive at the primary outputs before the appropriate falling edge, several clock cycles later. At the level-triggered latch L1, the data may depart at any time while the clock is high. A data signal in this circuit is allowed precisely two clock periods to reach the primary output from the primary input.

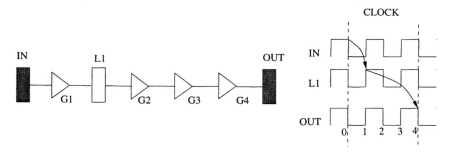

Figure 7.4.   Example illustrating the advantages of using level-sensitive latches.

We will now use this example to demonstrate the advantage of using level clocked circuits over edge-triggered circuits. For simplicity, we will assume zero setup and hold times here. Consider the operation of the circuit under a clock period of 2 units; notice that the path delay between latch L1 and the output is more than the clock period. However, the circuit works correctly due to the transparent nature of the latches. As shown in the figure, the data departs from the IN node at time 0, arrives at and departs from the latch L1 at time 1, and is latched at the output at time 4, which corresponds to the onset of the second clock edge. In contrast, if L1 were an edge-triggered FF, then a clock period of 2 units would have been untenable, since the clock period would correspond to the largest combinational block delay, which implies that the minimum possible clock period would have been 3 units. This practice of using the active period of the clock in a level-clocked circuit to allow combinational paths to have a delay of more than the clock period is referred to variously as either *cycle borrowing*, *cycle stealing*, *slack stealing*, or *time borrowing*.

For edge-triggered circuits, the insulating nature of the memory elements leads to the requirement that the delays through all combinational logic paths must be less than the clock period, with an allowance for setup and hold time constraints. Therefore, timing constraints need only be enforced between FF's connected by a purely combinational path. For level-clocked circuits, due to cycle borrowing, the delay through a combinational logic path may be longer than one clock cycle, as long as it is compensated by shorter path delays in subsequent cycles.

To ensure that the extra delay is compensated, we must enforce timing constraints between a latch and every other latch reachable from it via a path that traverses combinational logic, and possibly multiple latches.

Example. Consider a linear N stage acyclic pipeline with $N + 1$ memory elements $(m_0, m_1 \ldots m_N)$. If these memory elements were edge-triggered FF's, then we would need only N timing constraints (from the path $m_i \rightsquigarrow m_{i+1}$, $0 \leq i \leq N$). However, if these memory elements were to be level sensitive latches, then we would need $N \cdot (N + 1)/2$ timing constraints ($m_i \rightsquigarrow m_j \; \forall j > i$ and $0 \leq i \leq N$) to check the correctness of multicycle paths. In the presence of feedback paths, the timing analysis of level-clocked circuits would become even more complex.

As will soon be shown, some of this complexity can be reduced by the introduction of an appropriate set of intermediate variables.

7.3.2   Clock models

Single-phase and multiphase clocking.   The most conventional form of clocking uses a single-phase clock, where every rising [falling] clock edge is perfectly aligned with every other rising [falling] clock edge, assuming that no skews are introduced by the clock distribution network. Some high-performance circuits may, instead, use multiphase clocking; in a k-phase clock, the k phases typically have the same clock period, but are staggered from each other by a fixed time delay. Each phase consists of two intervals: an active interval during which the latches are enabled, and a passive interval when they are disabled; these typically correspond to the interval when the clock signal is high and low, respectively.

Depending on the way the clock scheme is designed, the phases may be overlapping or nonoverlapping, i.e., their active times may or may not simultaneously intersect in time with any other phase. Clocking schemes may be designed to be symmetric, which implies that all phases have an equal active time, and the time difference between the rising clock edges for successive phases i and j are all equal; a clock scheme that does not satisfy this property is said to be asymmetric.

Although it is not essential for correct operation, if one were to traverse any path in the circuit in the direction of signal propagation, one would typically find that a latch of phase i would be followed by a latch of phase $(i+1) \bmod k$; if so, the circuit is referred to as a well-formed circuit.

The SMO clock model.   This clock model, developed by Sakallah, Mudge and Olukotun [SMO92], and named after the initials of its authors, presents a complete description of a multiphase clock signal; this model will be used later in this chapter to describe the set of timing constraints for a level-clocked circuit, as derived by their work.

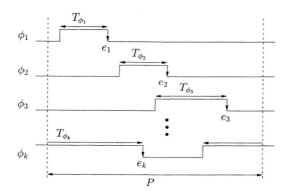

Figure 7.5.   A multiphase overlapping clocking scheme with k phase, with a period of P, and for each phase i, a falling edge time of e_i with respect to the global time reference and an active time of T_{ϕ_i} [SMO92].

A k-phase clock, illustrated in Figure 7.5, is a set of k periodic signals, $\Phi = \{\phi_1 \ldots \phi_k\}$, where $\phi_i, 1 \le i \le k$ is referred to as phase i of the clock Φ. All of the ϕ_i's have the same clock period, P. Each phase, i, has an active interval of duration T_{ϕ_i} and a passive interval of duration $(P - T_{\phi_i})$, corresponding to a duty cycle of $\frac{T_{\phi_i}}{P}$. Let Ψ denote the set of all latches. Each latch, $j \in \Psi$, is clocked by exactly one phase of the clock Φ, which is denoted by $p(j)$. All of the latches controlled by a clock phase are enabled during the active interval and disabled during the passive interval. Without loss of generality, it will be assumed in this discussion that the active interval corresponds to the clock being at logic value 1, while in the passive interval, the clock is at logic 0. The term "clocking scheme" is used to indicate the relative locations of the clock transition edges, and the duty cycles of the individual phases. Thus, a clocking scheme together with a clock period, P, define a clock schedule, Φ.

Associated with each phase, i, is a *local time zone*, shown in Figure 7.6, such that the passive interval starts at time 0, the enabling edge occurs at time $(P - T_{\phi_i})$, and the latching edge occurs at time P. Phases are ordered so that $e_1 \le e_2 \ldots \le e_{k-1} \le e_k = P$, and are numbered modulo-k, i.e., $\phi_{k+1} = \phi_1$ and

$\phi_{1-1} = \phi_k$. A global time reference is maintained as a general reference, and e_i denotes the time when the phase ϕ_i ends, relative to this global reference.

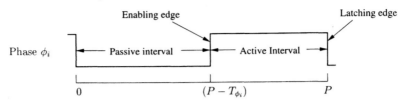

Figure 7.6.   Phase i of a k-phase clock with all times relative to the local time zone.

The phase shift operator, $E_{i,j}$, illustrated with the help of Figure 7.7, is defined as follows:

$$E_{i,j} = \begin{cases} (e_j - e_i) & \text{for } i < j \\ (P + e_j - e_i) & \text{for } i \geq j \end{cases} \tag{7.3}$$

Note that $E_{i,j}$ always takes on positive values, and acts as a readjustment factor in going from one time zone to another. Specifically, when $E_{i,j}$ is subtracted from a time point in the current time zone of ϕ_i, it changes the frame of reference to the local time zone of ϕ_j.

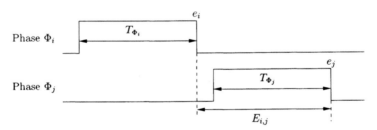

Figure 7.7.   Illustration of the phase shift operator, $E_{i,j}$.

The clock is required to satisfy the following constraints:

$$T_{\phi_i} \leq P \tag{7.4}$$
$$0 \leq e_i \leq P \tag{7.5}$$
$$e_i \leq e_{i+1} \tag{7.6}$$
$$P \geq 0 \tag{7.7}$$
$$T_{\phi_i} \geq 0 \tag{7.8}$$
$$E_{i,j} \geq T_{\phi_j} \tag{7.9}$$

The constraints (7.4), (7.5), (7.7) and (7.8) follow from the definition of the clock. The constraint (7.6) maintains an ordering of the phases, and (7.9) ensures that the phases are nonoverlapping. The last of these may be relaxed in case of an overlapping clock.

7.3.3   Timing constraints for level-clocked circuits

We now enumerate the set of timing constraints that dictate the correct operation of a level-clocked circuit. As before, the parameters $T_{s,i}$ and $T_{h,i}$ represent the setup and hold times, respectively, of the ith latch, and we assume δ_i and Δ_i to be the minimum and maximum delays through each latch. As defined earlier, for each pair of latches, i and j, connected by a combinational path $i \rightsquigarrow j$, the maximum and minimum path delays are denoted by $\overline{d}(i, j)$ and $\underline{d}(i, j)$, respectively.

The data input to each latch, i, has an associated arrival time that lies in the range between the earliest time, a_i, and the latest time, A_i. Similarly, the earliest and latest departure time of the data are denoted by d_i and D_i, respectively. All of these times are with respect to the local time zone of the latch.

Due to the transparent nature of the latches, a signal can depart from latch i at any time during the active interval of the phase $p(i)$, i.e., between time $P - T_{\phi_{p(i)}}$ and P. In other words,

$$P - T_{\phi_{p(i)}} \leq d_i \leq D_i \leq P \tag{7.10}$$

However, a signal cannot depart from a latch before it has arrived at that latch, i.e.,

$$A_i \leq D_i \tag{7.11}$$
$$a_i \leq d_i \tag{7.12}$$

We can use the definitions of D_i and d_i to obtain the relations

$$D_i = \max\{A_i, P - T_{\phi_{p(i)}}\} \tag{7.13}$$
$$d_i = \max\{a_i, P - T_{\phi_{p(i)}}\} \tag{7.14}$$

The arrival time at a latch, j, of a signal departing from another latch, i, connected by one or more purely combinational paths, $i \rightsquigarrow j$, must satisfy the following relations

$$D_i + \Delta_i + \overline{d}(i, j) - E_{p(i),p(j)} \leq A_j \tag{7.15}$$
$$d_i + \delta_i + \underline{d}(i, j) - E_{p(i),p(j)} \geq a_j \tag{7.16}$$

where $E_{p(i),p(j)}$ is the phase shift operator that translates a time in phase $p(j)$ into the local time in phase $p(i)$.

In addition, the latest arrival time must be at least one setup time before the falling edge of the clock, i.e., $A_i \leq P - T_{\phi_{p(i)}}$, and the earliest arrival time must occur no sooner than a hold time after the rising edge of the clock, i.e., $a_i \geq T_{h,i}$.

Combining all of these relations, we can obtain the timing constraints that must be imposed for a level clocked circuit to be correctly clocked as

$$A_j = \max_{i \leadsto j; i,j \in \Psi} \{D_i + \Delta_i + \overline{d}(i,j) - E_{p(i),p(j)}\} \tag{7.17}$$

$$D_i = \max(A_i, P - T_{\phi_{p(i)}}) \, \forall \, i \in \Psi \tag{7.18}$$

$$A_i \leq P - T_{s,i} \, \forall \, i \in \Psi \tag{7.19}$$

$$a_j = \min_{i \leadsto j; i,j \in \Psi} \{d_i + \delta_i + \underline{d}(i,j) - E_{p(i),p(j)}\} \tag{7.20}$$

$$d_i = \max(a_i, P - T_{\phi_{p(i)}}) \, \forall \, i \in \Psi \tag{7.21}$$

$$a_i \geq T_{h,i} \, \forall \, i \in \Psi \tag{7.22}$$

A critical path of a circuit corresponds to the path corresponding to a sequence of constraints associated with the gates and latches on the path, such that each of these constraints is an active or violated constraint (an active constraint corresponds to a $\geq$ or $\leq$ inequality that is just satisfied, i.e., is an equality). A critical path in a level-clocked circuit would typically traverse several combinational and sequential circuit elements. The term "critical path of a circuit" is often used colloquially to refer to the path with the largest violation, though sometimes, this term may also be used to refer to any violating path.

Critical paths may be classified into three categories:

Critical long paths are associated with long path constraints, or setup time constraints. Any increase in the delay along the path will worsen the slack or violation along that path.

Critical short paths are those that are associated with short path or hold time constraints. These paths are sensitive to a reduction in the delay along the path; any such reduction is liable to worsen the slack or violation along the path.

Critical cycles correspond to cyclic critical paths in the graph along which the constraints are active or violated.

For correct operation of a circuit, timing violations along all critical paths must be resolved. A detailed treatment of this subject is provided in [BSM95].

Timing verification.   Several procedures for timing analysis have been proposed, based on the SMO constraints, including [SMO92, BSM95, SS92, Szy92, LTW94]. In this section, we provide a brief synopsis of this work; the reader is referred to these papers for further details. The procedure *checkTc* [SMO90] proposed an iterative procedure for resolving the timing constraints. It was shown in [SS92] that the system of equations can be solved by relaxation in polynomial time in the number of latches, until a fixed point is obtained. It was also shown that the original system of equations in the SMO formulation have a unique solution unless the circuit operates at the minimum clock period. The following algorithm, SIMPLE-RELAX, shows an outline of the procedure:

```
Algorithm TIMING_VERIFY
/* initialize the latch arrival times */
for all latches i
    APrev[i] = aPrev[i] = −∞;
/* Iterate the evaluation of the departure and arrival time */
/* equations until convergence or a maximum of |L| iterations, */
/* where |L| is the number of latches.  */
iter = 0;
repeat
    iter++;
    /* Update the latch departure time based on the latch */
    /* Arrival times are computed in the previous iteration */
    for i = 1 to |L| {
        D[i] = max(APrev[i],P − T_{p(i)});
        d[i] = max(aPrev[i],P − T_{p(i)});
    }
    /* Update the latch arrival time based on the just-computed */
    /* latch departure times */
    for j = 1 to |L| {
        APrev[j] = A[j]; aprev[j] = a[j];
        A[j] = max_{i⤳j} (D[i] + Δ[i] + d̄(i,j) − E(p(i),p(j)));
        a[j] = min_{i⤳j} (d[i] + δ[i] + d̲(i,j) − E(p(i),p(j)));
    }
until ((A[i] == APrev[i]) && (a[i] == aPrev[i])) || (iter > |L| − 1);

/* Check and record setup and hold time violations */
for i = 1 to |L| {
    SetupVio[i] = (A[i] > P − T_{s,i});
    HoldVio[i] = (a[i] < T_{h,i});
};
```

The procedure is essentially a variant of the Bellman-Ford algorithm [CLR90], which will be described in detail in Section 9.3.4. In case the constraint set has some slacks, as is almost always the case, a general algorithm that finds values of each a_i, A_i, d_i and D_i that satisfies the "relaxed" constraints, (7.10), (7.11), (7.12), (7.15), and (7.16) is not obliged to report times that are physical. In contrast, this procedure finds the earliest arrival and departure times for the signals, corresponding to a physically meaningful solution.

Several variants of the procedure may be used to speed up the computation. For instance, by clipping off the departure times to physically realizable values (for example, setting $D_i = \min(\max(A_i, P − T_{p(i)}), P)$, the information detected is more relevant. If the signal does not arrive in time for the closing edge of the clock, it is signaled as a timing violation. This procedure serves to localize the effect of timing violations, and such a diagnostic provides useful input to procedures that are used to remedy such violations.

An alternative approach to this method utilizes the concept of back-edges [LTW94]. For a directed acyclic graph processed in topological order, the Bellman-Ford procedure (or the algorithm TIMING_VERIFY above) concludes in one iteration. It can be shown that if the graph has b back edges, b iterations are necessary. While the problem of identifying a minimal number of back-edges (i.e., the minimum feedback vertex set problem) is NP-hard, in real circuits, it is found that the number of back edges is small enough that almost any set is good enough to provide a good improvement over a naive implementation of the relaxation procedure above.

7.4 CLOCK SCHEDULE OPTIMIZATION FOR LEVEL-CLOCKED CIRCUITS

The clock schedule optimization problem is the problem of finding the minimum clock period, and the corresponding locations of the rising and falling edges of each phase so that the clocking scheme meets all timing constraints. Due to the freedom allowed in the arrival times for level-clocked circuits, this optimization can make a substantial difference to the behavior of the circuit.

The $minTc$ algorithm [SMO90] formulates the clock schedule optimization problem as a linear program. The problem is stated as follows:

$$\text{minimize} \qquad\qquad P \qquad\qquad\qquad (7.23)$$

subject to   Clock constraints in Relations (7.4) through (7.9)

Latch constraints in Relations (7.17) through (7.19)

After relaxing all max operators in relations (7.17) through (7.19) by replacing them by "$\geq$" inequalities, the problem maps on to a linear program whose minimum objective function value can be proven to be the same as that of the original problem.

The work in [Szy92] developed techniques to reduce the number of long path constraints in the linear program. A principal idea was the concept of *relevant paths*. Before defining this concept, we observe that if $r(p)$ denotes the number of latches on a path $p : e_i \rightsquigarrow e_j$, and $w(p)$ is the delay along the path (with allowances for setup times), then it must be true that

$$w(p) \leq r(p) \cdot P + E_{p(i),p(j)} \qquad\qquad (7.24)$$

A path, p, is said to be relevant for period P if for any other path, p', with the same endpoints as p, $w(p') - r(p') \cdot P < w(p) - r(p) \cdot P$ if $r(p') < r(p)$, and $w(p') \leq w(p)$ if $r(p') = r(p)$. It was shown that in order to check the feasibility of a clock schedule, it suffices to check if the constraints on the set of relevant paths are satisfied; the intuition behind this is that satisfying the relevant constraint will ensure that other constraints that were not relevant will be forced to satisfy Relation (7.24).

The work in [Szy92] uses this pruning to reduce the constraint set before feeding the resulting optimization problem to an LP solver. Subsequently, graph based algorithms for solving the k-phase clock scheduling problem were presented in [SBS92]; graph-based algorithms for clock schedule optimization for two-phase circuitry have also been presented in [ILP92]. For a fixed clock period, the timing constraints can be represented as a set of difference constraints of the type $x_i - x_j \leq c$, where the x_i's are the variables in the problem and c is a constant. This set of constraints can be represented by a constraint graph, and the longest path in the constraint graph yields a solution to the set of difference constraints, if one exists. If the clock period is infeasible, no such solution exists, and this is determined by a procedure such as the Bellman-Ford algorithm. Therefore, a binary search over the clock period is used to find the minimum value of the clock period at which the constraint graph has a solution; this procedure works since the search space is proven to be convex in P^2. The complexity of this algorithm is approximately $O(|\Psi|k^2 \log_2 |\Psi|)$, where $|\Psi|$ is the cardinality of the set, Ψ, of latches in the circuit, and k is the number of phases of the clock [SBS92].

A vital contribution of [Szy92], subsequently used in many other papers such as [SR94, SD96, MS98b], is the development of the predecessor heuristic to update the value of the search boundary on the detection of a positive cycle in the constraint graph; this is described in more detail in Section 9.3.5.

7.5   TIMING ANALYSIS OF DOMINO LOGIC

7.5.1   Introduction

Domino logic[3] is a popular choice of design style for implementing high speed logic designs since it has the advantage of low area and high speed. In this section, we will discuss techniques used for the timing analysis of domino logic.

Strictly speaking, our treatment here covers combinational circuits implemented using domino logic, but for various reasons related to the close relation of the clocking scheme with the operation of domino circuits, we treat this aspect of timing analysis in this chapter, rather than the one on combinational timing analysis.

The basics of domino gate operation may be explained with the help of a representative domino gate configuration is shown in Figure 7.8. When the clock input is low, the gate precharges, charging the dynamic node d to logic 1. In the next half-cycle of the clock when it goes high, the domino gate evaluates, i.e., the dynamic node either discharges or retains the precharged state, depending on the values of the input signals. The two-step mode of operation with a precharge and an evaluate phase causes the timing relationships in domino logic to be more complex than those for static logic.

The clock input to the domino gate is shown in Figure 7.8. The precharge phase begins at $T_{c,f}$ and continues until $T_{c,r}$, and the evaluate phase begins at that time and ends at time $T_{c,f} + P$ where P is the period of the clock signal feeding the domino gate. The reference time $t = 0$ is set with respect to the clock signal at the primary inputs. If more than one clock signal is used, any one of them may be used as a reference, and the transition times of the other clocks may be expressed according to the reference.

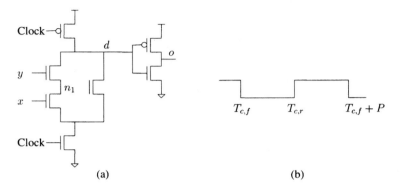

Figure 7.8.   (a) A typical domino circuit and (b) the clock waveform.

7.5.2 Timing constraints

The node timing constraints for domino logic may be expressed in terms of the signal arrival times and the clock arrival time. Note that in case of multiple clocks for the domino logic, the clock signal c should be set to be the clock signal that feeds the gate that is currently under consideration. The set of constraints were first presented in [VCL+96] and subsequently modified in [VMS96, ZS98]; the following presentation is the most complete of all and is taken from [ZS98].

(i) The first set of constraints addresses the arrival time of a falling transition at any data input of a domino gate. Any such falling event should meet the setup-time requirement to the rising edge of the evaluate clock to ensure that the dynamic node is not inadvertently discharged by a late arriving signal. If $T_f(in)$ refers to the falling event time of the input node, then it is required that

$$T_f(in) \leq T_{c,r} - T_{setup} \tag{7.25}$$

where the setup time T_{setup} is a constant that acts as a safety margin.

(ii) The second set of constraints is related to the arrival time of a rising transition at the output of the domino gate, and is presented here in terms of the arrival time of rising transitions at each of the input nodes. The constraint may be stated as follows: the rising event of the output node of the domino gate must be completed before the falling edge of evaluate clock. If $T_r(out)$ refers to the rising event time at the output node, then the circuit operates correctly only if

$$T_r(out) \leq T_{c,f} + P \tag{7.26}$$

In other words, before the beginning of the precharge for next cycle, the correct evaluation result must have traveled to the output node.

For example, in Figure 7.8, the rising event of the output node of domino gate with output o must satisfy (7.26). Therefore,

$$T_r(o) = \max((T_r(x) + D_f(x,d), T_r(y) + D_f(y,d), T_r(z) + D_f(z,d),$$
$$T_{c,r} + D_f(c,d))) + D_r(d,o) \quad (7.27)$$

where

$T_r(x), T_r(y), T_r(z)$ are the rising event times at inputs x, y and z, respectively
$D_f(i,d)$ represents the delay of a falling transition at the dynamic
node d due to a rising transition at input $i \in \{x, y, z\}$
$D_r(d,o)$ represents the rise delay of the inverter feeding the gate output node o
$D_f(c,d)$ is the delay from the clock node c to the dynamic node d

Therefore for $i \in \{x, y, z\}$,

$$D_f(i,d) + D_r(d,o) - P \leq T_{c,f} - T_r(i) \quad (7.28)$$
$$D_f(i,d) + D_r(d,o) - P \leq T_{c,f} - T_{c,r} \quad (7.29)$$

The relation (7.28) corresponds to the requirement that the rising edge of each input should appear in time for the falling edge of the evaluate clock so as to allow sufficient time for the output to be discharged.

The relation (7.29) ensures that the pulse width of the evaluate clock is sufficient for pulling down the output node when the last transistor to switch is the lowermost one, connected to the clock node.

(iii) The third set of constraints addresses the timing requirements on rise transitions at the dynamic node d. The rising event d of the domino gate must be completed before the rising edge of the evaluation clock, i.e.,

$$T_r(d) \leq T_{c,r} \quad (7.30)$$

If the rise time of the dynamic node through the p-transistor fed by the clock is denoted by $D_r(c,d)$, then the rising event time can be expressed as:

$$T_r(d) = T_{c,f} + D_r(c,d) \quad (7.31)$$

This leads to the constraint given by

$$D_r(c,d) \leq T_{c,r} - T_{c,f} \quad (7.32)$$

This implies that the pulse width of precharge must be capable of pulling up the output node.

Note that unlike (ii) above, the delay to only node d is considered here, and not to the output node o. Note that the constraint to node o is very loose and is satisfied except in the most pathological cases. There are two possibilities:

- if the evaluate phase causes the dynamic node to go low, then the completion of the fall transition at o during precharge is irrelevant.

- if the evaluate phase causes the dynamic node to remain high, then it will remain so until the end of the precharge phase in the next clock cycle, and the node o correspondingly has a long time (equal to $P + T_{c,r}$) to discharge, and therefore the constraint is very loose.

In most cases, the overriding timing constraint will be due to some other constraint, such as that corresponding to a gate at which the signal o is being utilized.

(iv) In the previous clock cycle, a falling input data line may not go down to logic 0 but must be held at the logic 1 level until the output transition in that cycle has been completed. In other words, it is required that

$$T_f(in) \geq T_{c,r} - P + \max_{\forall in \in I_{domino}} (D_f(in, d) + D_r(d, o)) \qquad (7.33)$$

The term $D_r(d, o)$ is conservatively added here, since the fact that node d has reached the 50% point as it transitions to logic 0 does not, in general, guarantee that the output inverter will charge up to 1. Although the output inverter is often sized in a skewed manner, so that if d is at the 50% point, then the output will be well on its way to discharging, there may be issues related to short-circuit current that require this stronger criterion.

(v) For any rising input data line, the hold constraint states that a transition should occur only after $T_{c,f}$, i.e.,

$$T_r(in) \geq T_{c,f} \qquad (7.34)$$

These constraints are similar to the short path constraint in static timing analysis.

7.5.3   Timing analysis

The timing analysis procedure described here is based on the CPM procedure. In mixed static-domino circuits, the static elements can easily be represented by nodes in the CPM graph. Dynamic nodes can also be represented as nodes in the CPM graph by representing output rise and fall times by expressions similar to Equation (7.27), and then using the relaxed version of the constraint, replacing each "max" by an inequality. This may be intuitively seen by observing that the relations in Section 7.5.2 can be written in terms of difference constraints [CLR90].

The determination of rising and falling event arrival times at the output node of a domino gate is similar to the CPM computation for static gates. A major difference is that the rising event at the dynamic node is related only to the falling edge of the domino clock and is independent of the other input nodes. This effect is captured by setting the value of D_r from each input node of the domino gate to the output node as $-\infty$. The domino clock input node is treated in the same way as any primary input node, and the rising or the falling edge of the clock provide the corresponding event times for the clock node. In case the clock input to the domino gate is the output of another gate, its event time is given by the maximum of all gate input event arrival times, plus the transition delay of the gate.

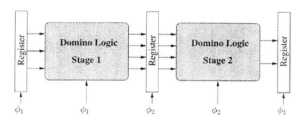

Figure 7.9.   A two-stage domino pipeline using a two-phase clock.

7.5.4 Sequential circuits using domino logic

Typical domino circuits use multiphase clocks. For a simple two-phase domino pipeline shown in Figure 7.9, alternate phases are clocked by different clock phases. Therefore, when the first stage goes through precharge, the second stage evaluates, and vice versa. This is clearly better than a situation where both stages precharge at the same time and evaluate simultaneously, since it permits some operations to overlap. A good source to explore multiphase domino logic, and the design of skew-tolerant domino circuits, in greater detail is [HH97].

7.6 SUMMARY

In this chapter, we have overviewed the essential constraints that must be verified during sequential timing analysis. For edge-triggered circuits, each combinational stage can be treated independently, but for level-clocked circuits, where multicycle paths may exist, a more involved analysis is necessary. Finally, we have reviewed techniques for the timing analysis of domino logic circuits.

It is often useful to arrive at a "black box" characterization of a block, for instance, an intellectual property (IP) block. Algorithms for this purpose have been presented in [VPMS97, YMP+01], and alternative models for this representation are examined in [DMS+02].

Notes

1. Recall, from Chapter 5, that any cycles in combinational circuits are broken, so that it is assumed here that any combinational circuit must be acyclic.

2. This basic procedure is widely used in many other algorithms for clock skew optimization and retiming, presented in Chapters 9 and 10, respectively, and the reader is invited to read those sections for a fuller understanding.

3. Although domino logic is often used to implement combinational elements, we discuss it in this chapter since this explanation requires some prior discussion of clocking issues.

8 TRANSISTOR-LEVEL COMBINATIONAL TIMING OPTIMIZATION

8.1 INTRODUCTION

A typical digital integrated circuit consists of combinational logic blocks that lie between memory elements that are clocked by system clock signals. For such a circuit to obey long path and short path timing constraints, the delays of the blocks should be adjusted so that they satisfy these constraints. As seen in Chapter 7, this is easier for edge-triggered circuits than for level-clocked circuits since for the former, the delay constraints can be applied separately to each combinational block; for the latter, the existence of multicycle paths makes the problem more involved, but it is still tractable. In our discussion in this chapter, we will focus primarily on the timing optimization of edge-triggered circuits since they are more widely used; extensions to level-clocked circuits are primarily based on very similar methods.

For an edge-triggered circuit, the timing constraints dictate that valid signals must be produced at each output latch of a combinational block before any transition in the signal clocking the latch. In other words, the worst-case input-output delay of each combinational stage must be restricted to be below a certain specification. This may be achieved in various parts of the design cycle, but we will primarily focus on the role of back-end design tools here.

Timing optimization may be performed at various steps of the design cycle. The technology-independent synthesis step [De 94] typically uses simple metrics for timing: for instance, the delay of a circuit may be considered to be the number of levels of logic. Truer delay metrics are used in the technology-dependent synthesis, or technology mapping, step, where cell-level delay models are often used for timing optimization [Keu87, CP95, HWMS03]. In the physical design step, place and route algorithms typically attempt to perform timing optimization, sometimes using metrics that are correlated with the delay, such as wire length minimization, and sometimes using delay metrics directly. Due to the complexities involved in placement, such metrics are necessarily simple. As the physical layout gains more concreteness, more accurate models may be applied. In case of global routing, where the choice of a route and the use of buffers can significantly impact the wire delay, more complex delay models are employed for timing optimization [van90, ADQ99, HS00].

The breakdown of the design cycle into synthesis, physical design, and transistor level optimization steps is somewhat artificial, as these steps are not entirely independent of each other. A growing realization of this has led to significant thrusts in physical synthesis, which focus on the interaction between synthesis and physical design. A typical mode of operation in physical synthesis may involve working with a physical prototype for the placement upon which the wire length models that drive technology mapping are based [Dai01].

Transistor level design offers one of the best opportunities for the use of more detailed timing models, since the degrees of freedom are greatly narrowed down by this stage. Several options for transistor-level timing optimizations are available. One powerful tool for timing optimization is the use of transistor sizing: instead of using all minimum-sized or all uniform-sized transistors, this optimization determines the sizes that optimize an objective function such as area or power. Another optimization that is particularly useful in delay reduction at the transistor level is the use of dual threshold voltages, whereby a tradeoff between the leakage current and the delay may be arrived at by selectively mixing low V_t transistors with high V_t transistors in the design. In this chapter, we will overview these transistor-level optimizations, which will build upon the timing analysis techniques that we have surveyed in the previous chapters.

8.2 TRANSISTOR SIZING

8.2.1 Introduction

Transistor sizing involves the determination of an optimal set of transistor sizes for a circuit, typically using objective or constraint functions that are related to timing, power and area. This problem is frequently posed as a nonlinear programming problem, and several approaches have been proposed in the past. The "size" of a transistor typically refers to its channel width (unless otherwise specified), since the channel lengths of MOS transistors in a digital circuit are generally uniform[1]. In coarse terms, the circuit delay can usually be reduced by increasing the sizes of certain transistors in the circuit (although indiscriminate increases are not very useful since the actual relation between transistor sizes and the circuit delay is more complex). Hence, making the circuit faster usually entails the penalty of increased circuit area. The area-delay trade-off involved here is, in essence, the problem of transistor size optimization.

A simple example from [Hed87] serves to illustrate how the delay varies with transistor sizes. Consider the chain of three CMOS inverters shown in Figure 8.1(a). For simplicity, assume that all of the transistors have the same channel width. Let the width of both the n-type and p-type transistors in gate G_2 be w_2, and let D be the total delay through the three gates.

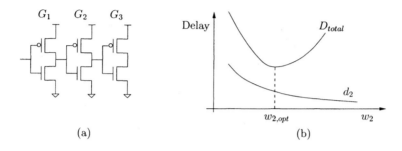

(a)   (b)

Figure 8.1.   (a) A chain of three inverters, and (b) the effect of transistor sizes on delay for the three-inverter chain.

Consider the effect of increasing w_2, while keeping the size of the transistors in gates G_1 and G_3 fixed. This causes the magnitude of the output current of gate G_2 to increase. Thus the time required, d_2, for gate G_2 to drive its output signal will decrease monotonically, as shown in Figure 8.1(b). However, increasing w_2 also increases the capacitive load on the output of gate G_1, thus slowing down the output transition of the first gate. Beyond a certain optimal point $w_2 = A$, at which the total delay, D, is minimized, the delay begins to increase with respect to w_2.

This example illustrates the tradeoff involved in increasing a transistor size: while the drive of the transistor is increased, so is the load that it offers to the previous stage. The real problem of transistor sizing is considerably more complex since it involves a larger number of variable, and delays along a larger number of paths, as the sizes of all transistors in a circuit are optimized.

We will now consider various formulations of this problem, followed by a set of solution techniques. Specifically, we will survey the heuristic approach in TI-LOS, a more formal convex programming method in iCONTRAST, Lagrangian multiplier methods, and timing budget based procedures. Other methods using linear programming have also been utilized [BJ90], with delay models that are piecewise linear functions of the transistor sizes. These will be covered in the discussion of the combined skew and sizing optimization method of [CSH95] in Chapter 9.

8.2.2 Formulation of the optimization problem

A basic formulation. The transistor sizing problem for a combinational block in an edge-triggered circuit is most commonly formulated as

$$\text{minimize} \quad Area = \sum_{i=1}^{n} \alpha_i x_i$$
$$\text{subject to} \quad Delay \leq T_{spec}. \tag{8.1}$$

where x_i is the size of the ith transistor (out of n total transistors) in the circuit. The objective function approximates the area of a circuit as a weighted sum of transistor sizes, a metric that is correlated with the circuit area, but not an exact measure. An accurate metric for the area occupied by a circuit requires consideration of the layout arrangement, the cell structures and the design style. This is unlikely to yield a well-behaved functional form, and therefore the above approximation is widely used since it possesses the properties of continuity and differentiability that are desirable from the point of view of a general nonlinear optimizer. Consequently, from the point of view of transistor sizing, the approximate area function in the objective is considered adequate.

The constraint function for edge-triggered circuits simply states that the delay of each combinational block should not exceed a number, T_{spec}, which depends on the clock period, the setup times, and the skew tolerances. Strictly speaking, a hold time constraint should also be incorporated, but this is often taken care of in the design methodology in a separate step from transistor sizing. For level-clocked circuits, the delay constraints are more complex and must take into account the possibility of multicycle paths, as described in Section 7.3.

In addition to Equation (8.1), several other formulations may be useful. These include

- Minimizing $Area \cdot Delay^{\gamma}$, for some exponent γ.

- Minimizing the dynamic, short-circuit, or active/standby leakage power under an area specification, $Area \leq A_{spec}$.

A formulation that enumerates all constraints. In Equation (8.1), the constraint $Delay < T_{spec}$ implies that the delay on *every* combinational path in the circuit must be less than T_{spec} (for an edge-triggered circuit). The number of paths in a circuit could be exponential in the number of gates in a circuit, and therefore, any optimizer that requires these constraints to be enumerated literally experiences great difficulty with such a formulation. For this reason, the formulation in (8.1) is used by optimizers such as TILOS and iCONTRAST that do not require constraint enumeration.

For optimizers where the constraints must be explicitly listed, a method suggested in [Mar86, MG87, MG86] employs intermediate variables to reduce the number of delay constraints from exponential to linear in the circuit size. This is achieved as follows. As in the CPM method, the circuit is modeled by a graph, $G(V, E)$, where V is the set of nodes (corresponding to gates) in G, and E is the set of edges (corresponding to connections among gates). If m_i represents the worst-case arrival time of a signal at the output of gate i, then for each gate, the delay constraint is expressed as

$$m_i + d_j \leq m_j. \tag{8.2}$$

where gate $j \in$ fanout(gate i), and d_j is the delay of gate j. Therefore, the transistor sizing problem can be framed as [CCW99]

$$
\begin{aligned}
\text{minimize} \quad & \textstyle\sum_{i=1}^{n} \alpha_i x_i \\
\text{subject to} \quad & m_i \leq M_0 && \forall \ i \in \text{primary outputs} \\
& m_i \geq M_i && \forall \ i \in \text{primary inputs} \\
& m_i + d_j \leq m_j && \forall \ \text{inputs } j \text{ of all gates } i \\
& L_i \leq x_i \leq U_i && \forall \ \text{transistors } i
\end{aligned}
\tag{8.3}
$$

Here, M_0 for primary output i is the arrival time specification at the output, and M_i at primary input i is the specified arrival time at the input (in the most common case where a combinational block between flip-flops is being considered, $M_i = 0$ for all primary inputs i). For all other gates i, m_i is the arrival time at the output and d_j is the delay of the gate. The rising and falling arrival times may be considered separately within this framework. As before, α_i is a weight associated with the size of the ith transistor.

Thus, the number of delay constraints is reduced from a number that could, in the worst case, be exponential in $|V|$, to one that is linear in $|E|$, by the addition of $O(|V|)$ intermediate variables.

Practically, the number of constraints can be substantially reduced even over this, using a technique proposed in [VC99]. The essential idea of this approach is based on the observation that the introduction of the intermediate variables above does not *always* win over path enumeration. In cases where subpath enumeration reduces the number of constraints, some constraints could be combined, and the intermediate arrival time variables eliminated. As an example, consider a case of a chain of three inverters, as shown in Figure 8.1, and let us assume that the rise and fall delays are processed separately. The circuit has two paths, and if the required delay at the output is T_{spec}, then assuming that the inputs are available at time 0, the constraints can be written as

$$t_{r,1} + t_{f,2} + t_{r,3} \ \leq \ T_{spec} \tag{8.4}$$

$$t_{f,1} + t_{r,2} + t_{f,3} \ \leq \ T_{spec} \tag{8.5}$$

where $t_{r,i}$ and $t_{f,i}$ are, respectively, the rise and fall delays of gate G_i. On the other hand, if intermediate arrival time variables are used, the number of variables and constraints is each clearly larger:

$$t_{r,1} \leq m_{r,1} \qquad (8.6)$$

$$m_{r,1} + t_{f,2} \leq m_{f,2} \qquad (8.7)$$

$$m_{f_2} + t_{r,3} \leq T_{spec} \qquad (8.8)$$

$$t_{f,1} \leq m_{f,1} \qquad (8.9)$$

$$m_{f,1} + t_{r,2} \leq m_{r,2} \qquad (8.10)$$

$$m_{r_2} + t_{f,3} \leq T_{spec} \qquad (8.11)$$

where $m_{r,i}$ and $m_{f,i}$ are the intermediate variables that represent the rise and fall arrival times, respectively, at the output of gate $G_{i\ell}$

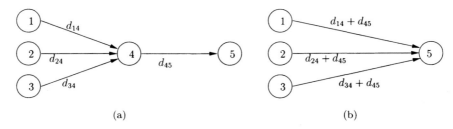

(a)   (b)

Figure 8.2. The principle behind reducing the number of constraints and intermediate variables in a timing graph. (a) shows an initial configuration and (b) its replacement by the elimination of the internal variable corresponding to the central node [VC99].

More generally, consider a timing graph substructure such as that shown in Figure 8.2(a). A direct introduction of the intermediate arrival time variables results in 1 rise and 1 fall constraint at each of nodes 1, 2 and 3; 3 rise and 3 fall constraints at node 4; and 1 rise and 1 fall constraint at node 5, for a total of 10 constraints. When the internal node is eliminated, the input-to-output constraints for this graph are shown in Figure 8.2(b): this corresponds to 3 rise time and 3 fall time constraints, for a total of 6 constraints. Generalizing this, for a substructure with m input nodes, n output nodes, and one internal node, the number of constraint using intermediate arrival time variables is $2(m + n + 1)$, with one arrival time variable for each of the m input nodes, the n output nodes and the internal node, and a factor of 2 corresponding to the fact that both rise and fall constraints must be listed. If the internal node is eliminated, the number of constraints reduces to $2mn$. The approach used in [VC99] uses this very criterion to identify a candidate node for elimination, and to eliminate it if $2mn \leq 2(m + n + 1)$.

An example of how this is applied is illustrated on a timing graph in Figure 8.3. The nodes s and t represent the source and sink vertex, respectively, and the nodes 1 through 16 correspond to combinational delays. The number of timing constraints using intermediate variables is equal to the number of edges (26), and the number of intermediate variables equals the number

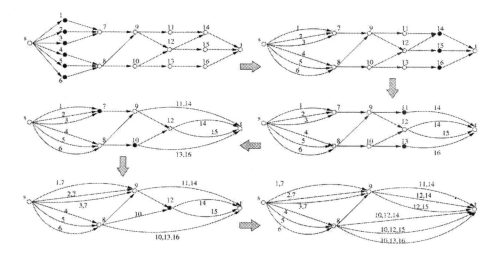

Figure 8.3. An example [VC99] to illustrate a reduction in the number of intermediate variables and constraints for a timing graph. In each step, the black vertices are eliminated and the timing constraints are updated.

of nodes (18). The first step removes the intermediate variables for nodes 1 through 6 to obtain the second graph, reducing the number of nodes to 12 and the number of edges to 20. Next, nodes 14, 15 and 16 are removed, and the number of nodes and edges reduces further to 9 and 15, respectively. By eliminating nodes 11 and 13, the next step reduces these numbers to 5 and 13, respectively. Subsequent steps remove nodes 7 and 10, followed by node 12, to obtain a final graph with just two internal nodes. Therefore, the net result of these transforms is that the number of intermediate variables is reduced from 18 to 4, while the number of constraints is halved, from 26 to 13.

On a set of industrial circuits, this method was shown to reduce the number of arrival time variables by nearly 90%, the total number of variables by about 25%, and the total number of timing constraints by nearly 40%.

8.3   THE TILOS ALGORITHM

The algorithm that was implemented in TILOS (TImed LOgic Synthesizer) [FD85, DFHS89, HSFK89] was the first to make use of the fact that the area and delay can be represented as posynomial functions (to be defined in Section 8.3.2) of the transistor sizes.

The TILOS algorithm is simple to implement on top of a static timing analyzer. The method is essentially a simple greedy heuristic that is surprisingly effective on many circuits. The basic philosophy is that for the problem statement in Equation (8.1), with all weights α_i set to 1, the objective function is minimized when all of the transistors are at the minimum size permitted by the technology, and hence this is used as the initial solution. An STA step is then carried out to check whether this initial solution satisfies the timing constraints or not; if so, the solution has been found, and if not, an iterative procedure is employed. In each iteration, the area is allowed to increase in such a way that the largest impact is made on delay reduction. To reduce the circuit delay, the area increase must necessarily be restricted to transistors on the critical path. The ratio $\frac{\partial Delay}{\partial Area}$ is a measure of the "bang per buck," and the transistor in a gate on the critical path that provides the largest such ratio is upsized. The STA step is repeated, and the iterations continue until the timing constraints have been met.

The core of the algorithm requires a timing analysis and a sensitivity computation that calculates the delay change when a transistor size is perturbed; this may be computed either using a sensitivity engine or by using finite differences and multiple calls to the STA engine. In each iteration, since only a minor change is made to the circuit (i.e., one transistor size is altered), incremental timing analysis techniques may be employed to reduce the overhead of timing calculation.

8.3.1   The TILOS delay model

We examine delay modeling in TILOS at the transistor, gate and circuit levels. At the transistor level, an RC model is used, with an approximated linear source-to-drain resistance, and capacitances associated with the source, drain and gate nodes. The resistance is inversely proportional to the transistor size, while the capacitances are directly proportional, with different constants of proportionality for the gate node and the source/drain nodes.

At the gate level, TILOS operates in the following manner. For each transistor in a pullup or pulldown network of a complex gate, the largest resistive path from the transistor to the gate output is computed, as well as the largest resistive path from the transistor to a supply rail. Thus, for each transistor, the network is transformed into an equivalent RC line corresponding to this path[2], and the Elmore time constant for this RC line is computed. This Elmore delay corresponds to the delay of the gate when the transition is caused by the transistor under consideration, and can be used to compute input-to-output pin delays for all input pins in the gate. At the circuit level, the CPM technique, described in Section 5.2.1, is used to find the circuit delay, and gate delays are computed on the fly as a gate is scheduled for processing.

8.3.2   Posynomial properties of the delay model

The circuit delay model above can be shown to fall into a class of real functions known as *posynomials* that have excellent properties that can be exploited during optimizations. A posynomial is a function p of a positive variable $\mathbf{x} \in R^n$ that has the form

$$p(\mathbf{x}) = \sum_j \gamma_j \prod_{i=1}^{n} x_i^{\alpha_{ij}} \qquad (8.12)$$

where the exponents $\alpha_{ij} \in \mathbf{R}$ and the coefficients $\gamma_j \in \mathbf{R}^+$. Simply stated, this means that a posynomial is similar to a polynomial, with the difference that each term must have a positive coefficient, and the exponents associated with the variables may be arbitrary real numbers that could be positive, negative, or zero. In the positive orthant in the $\mathbf{x}$ space, posynomial functions have the useful property that they can be mapped onto a convex function through an elementary variable transformation, $(x_i) = (e^{z_i})$ [Eck80].

The delay model discussed in Section 8.3.1 uses resistor models that are proportional to the inverse of some transistor sizes, and capacitor models that are directly proportional to some other transistor sizes, and uses RC products to compute gate delays. The delay on any path, computed using CPM, is therefore a sum of such RC products, and can easily be verified to use the following form of expressions:

$$D(\mathbf{x}) = \sum_{i,j=1}^{n} a_{ij} \frac{x_i}{x_j} + \sum_{i=1}^{n} \frac{b_i}{x_i} + K \qquad (8.13)$$

where $a_{ij}, b_i, K \in \mathbf{R}^+$ are constants and, $\mathbf{x} = [x_1, \cdots, x_n]$ is the vector of transistor sizes. This shows that the model expresses the path delay as a posynomial function of the transistor sizes; specifically these posynomials have exponents that belong to the set $\{-1,0,1\}$.

The delay of a combinational circuit is defined to be the maximum of all path delays in the circuit. Hence, it can be formulated as the maximum of posynomial functions. This is mapped by the variable transformation $(x_i) = (e^{z_i})$ onto a maximum of convex functions, which is also a convex function. The area function is linear in the x_i's and is therefore also a posynomial, and is thus transformed into a convex function by the same mapping. Therefore, the optimization problem defined in (8.1) is mapped to a *convex programming* problem, i.e., a problem of minimizing a convex function over a convex constraint set. A basic result in optimization theory states that for any such problem, any local minimum is also a global minimum.

8.3.3 The TILOS optimizer

The optimization algorithm used by TILOS assumes an initial solution where all transistors are at the minimum allowable size. For a general sequential circuit, TILOS first extracts the combinational subnetworks and their input-output timing requirements, and then performs the steps described in this section on each combinational subnetwork. Keeping this in mind, for the rest of this section we can assume, without loss of generality, that the circuit to be optimized is purely combinational.

In each iteration, a static timing analysis is performed on the circuit, which assigns two numbers to each electrical node: t_f, the latest fall transition time, and t_r, the latest rise transition time. This timing analysis is used to determine the critical path for the circuit. Let N be the primary output node on the critical path. The algorithm then walks backward along the critical path, starting from N. Whenever an output node of gate i is visited, TILOS examines the largest resistive path between V_{DD} and the output node [between ground and the output node] if the rise time [fall time] of i causes the timing failure at N. This includes

- The *critical transistor*, i.e., the transistor whose gate is on the critical path.

- The *supporting transistors*, i.e., transistors along the largest resistive path from the critical transistor to the power supply (V_{DD} or ground).

- The *blocking transistors*, i.e., transistors along the highest resistance path from the critical transistor to the logic gate output.

TILOS finds the sensitivity, which is the reduction in circuit delay per increment of transistor size, for each critical, blocking and supporting transistor. The size of the transistor with the greatest sensitivity is increased by multiplying it by a constant, BUMPSIZE, a user-settable parameter that defaults to 1.5. Practically, it is observed that smaller values of 1.01, 1.05 or 1.1 can provide better quality solutions, often without a very large CPU time penalty. The above process is repeated until

- all constraints are met, implying that a solution is found, or

- the minimum delay state has been passed, and any increase in transistor sizes would make it slower instead of faster, in which case TILOS cannot find a solution.

The reason for increasing the transistor size by the factor BUMPSIZE, rather than minimizing the delay along the critical path, is that such a minimization would not necessarily optimize the delay of the circuit, since another path may become critical instead; in such a case, the minimization would be overkill and may involve an excessively large and unnecessary area overhead. Instead, the delay along the current critical path is gradually reduced by this method, and at the point at which another path becomes critical, its delay is reduced instead, and so on.

Note that since in each iteration, exactly one transistor size is changed, the timing analysis method can employ incremental simulation techniques to update delay information from the previous iteration. This substantially reduces the amount of time spent by the algorithm in critical path detection.

The sensitivity calculation for the critical path can be carried out in a computationally efficient manner. When a transistor i has its size bumped up by the factor, BUMPSIZE, the delay of all gates on the path under consideration remains unaltered, except for

- the gate in which the transistor lies: since its driving power is increased, its delay decreases.

- the gate that drives this transistor: since it experiences a larger load, its delay increases.

By algebraically summing up these two changes, the change in the delay of the path is easily found[3].

8.4 TRANSISTOR SIZING USING CONVEX PROGRAMMING

The approach in [SRVK93] applies an efficient convex programming method [Vai89] to guarantee global optimization over the parameter space of all transistor sizes in a combinational subcircuit, thereby solving the optimization problem exactly.

The algorithm works in the space $\mathbf{R}^n$, where n is the number of transistors in the circuit. Each point in the space corresponds to a particular set of transistor sizes. The feasible region is the set of points (i.e., assignments of transistor sizes) that satisfies the delay constraints.

The algorithm works in the transform domain z_i under the transformation $(x_i) = (e^{z_i})$, and starts by bounding the convex domain by an initial polytope. A polytope is simply an n-dimensional generalization of a convex polygon, and can be represented by the intersection of a set of linear inequalities as

$$P = \{\mathbf{z} : A\mathbf{z} \geq \mathbf{b}\} \tag{8.14}$$

where A is a constant $n \times n$ coefficient matrix, and $\mathbf{b}$ is a constant n-vector. The initial polytope is typically a box described by the user-specified minimum and maximum values for each transistor size. By using a special cutting plane technique, the volume of this polytope is shrunk in each iteration, while ensuring that the optimal solution lies within the boundary of the reduced polytope. The iterative procedure stops when the volume of the polytope becomes sufficiently small.

Let $\mathbf{z}_{opt}$ be the solution to the optimization problem in the transformed z_i domain. The objective of the algorithm is to start with a large polytope that is guaranteed to contain $\mathbf{z}_{opt}$, and in each iteration, to shrink its volume while keeping $\mathbf{z}_{opt}$ within the polytope, until the polytope becomes sufficiently small. The algorithm proceeds as follows:

Step 1 An approximate center $\mathbf{z}_c$ for the polytope is found deep in its interior. This is achieved by minimizing a log-barrier function, and constitutes the most computationally intensive part of the approach; for details, the reader is referred to [SRVK93].

Step 2 An static timing analysis is performed to determine whether or not the transistor sizes corresponding to the center satisfy the timing constraints, i.e., whether $\mathbf{z}_c$ lies within the feasible region.

In case of infeasibility, the convexity of the constraints guarantee that it is possible to find a *separating hyperplane*

$$\mathbf{c}^T\mathbf{z} \geq \beta \tag{8.15}$$

passing through $\mathbf{z}_c$ that divides the polytope into two parts, such that the feasible region lies entirely on one side of it; the determination of this hyperplane only requires the gradient of the critical path delay since it is specified by

$$\mathbf{c} = -[\nabla D_{critpath}(\mathbf{z})]^T$$
$$\beta = \mathbf{c}^T \mathbf{z}_c$$

If the point $\mathbf{z}_c$ is feasible, then the convexity of the objective function implies that there exists a hyperplane that divides the polytope into two parts such that $\mathbf{z}_{opt}$ is contained in one of them; this time, the hyperplane is specified by the gradient of the objective function as

$$\mathbf{c} = -[\nabla Area(\mathbf{z})]^T$$
$$\beta = \mathbf{c}^T \mathbf{z}_c$$

Step 3 In either case, the constraint (8.15) is added to the current polytope to give a new polytope that has roughly half the original volume.

Step 4 The process is repeated until the polytope is sufficiently small.

We now illustrate the application of the algorithm to solve the problem

$$\text{minimize} \quad f(x_1, x_2)$$
$$\text{such that} \quad (x_1, x_2) \in S \qquad (8.16)$$

where S is a convex set and f is a convex function. The example problem is pictorially depicted in Figure 8.4(a), where the shaded region corresponds to the feasible region S, and the dotted lines show contours of constant value of the function f. The point $\mathbf{x}^*$ is the solution to this problem.

The expected solution region is first bounded by a closed initial polytope, which is a rectangle in two dimensions, as shown in Figure 8.4(a), and its center, $\mathbf{z}_c$, is found. We now determine whether or not $\mathbf{z}_c$ lies within the feasible region. In this case, it can be seen that $\mathbf{z}_c$ lies outside the feasible region. Hence, the gradient of the constraint function is used to construct a hyperplane through $\mathbf{z}_c$, such that the polytope is divided into two parts of roughly equal volume, one of which contains the solution $\mathbf{x}^*$. This is illustrated in Figure 8.4(b), where the shaded region corresponds to the updated polytope.

The process is repeated on the new smaller polytope. Its center lies inside the feasible region; hence, the gradient of the objective function is used to generate a hyperplane that further shrinks the size of the polytope, as shown in Figure 8.4(c). The result of another iteration is illustrated in Figure 8.4(d). The process continues until the polytope has been shrunk sufficiently according to a user-specified criterion.

8.5   LAGRANGIAN MULTIPLIER APPROACHES

The method of Lagrangian multipliers [Lue84] is a general method that can be used to solve nonlinear optimization problems, and has been applied to solve the transistor sizing problem.

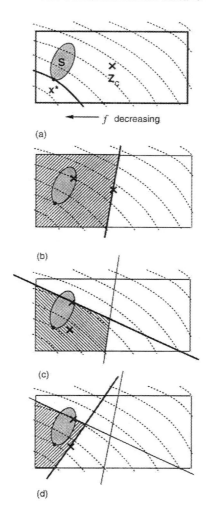

Figure 8.4.   An example illustrating the convex programming algorithm using in [SRVK93].

8.5.1   Early approaches

An early approach in [Cir87] minimizes the area of a single path subject to delay constraints on the path by formulating the problem using Lagrangians. Another approach in [Hed87] performs this optimization for an enumerated set of critical paths using a smoothing function to approximate the max function using a continuous function. However, any path based-method is liable to fail since the number of paths in a circuit can be exponential in the number of gates.

The approach in [Mar86, MG87, MG86] presents a solution to transistor size optimization by using Lagrangian multipliers, and employs intermediate variables to reduce the number of delay constraints from an exponential number to a number that is linear in the circuit size. The optimizer employs two variations of the Lagrange multiplier approach, the augmented Lagrangian algorithm and the projected Lagrangian algorithm [Lue84, Mar86]. The augmented Lagrangian algorithm uses a penalty term that helps to steer the solution towards the feasible region and, hence, has the desirable property of global convergence. The projected Lagrangian method, a quadratic approximation method that is similar to Newton's method, has a fast convergence rate. However, it is not globally convergent, and requires that the initial solution be close enough to the optimum solution in order to converge. COP starts with the augmented Lagrangian technique to take advantage of its global convergence. When the gradient of the Lagrangian becomes less than a certain small number, ϵ, it switches to the projected Lagrangian algorithm, to converge more quickly to the solution.

8.5.2 Lagrangian relaxation methods

The approach in [CCW99] presents a detailed analysis of the Lagrangian of Equation (8.3). The problem can be rewritten as

$$\text{minimize} \qquad L_\lambda(\mathbf{x}, \mathbf{m}) \qquad\qquad (8.17)$$
$$\text{subject to} \quad L_i \leq x_i \leq U_i \ \ \forall \ \text{gates} \ i$$

where

$$
\begin{aligned}
L_\lambda(\mathbf{x}, \mathbf{m}) \ = \ & \sum_{i=1}^{n} \alpha_i x_i + \textstyle\sum_{i \ \in \ \text{primary outputs}} \lambda_{i0}(m_i - M_0) + \\
& \textstyle\sum_{i \ \in \ \text{all inputs}} \lambda_{mi}(M_i - m_i) + \\
& \textstyle\sum_{\text{all other gates } i} \textstyle\sum_{j \ \in \ \text{inputs of } i} \lambda_{ji}(m_j + d_j - m_i)
\end{aligned}
$$

$$\qquad\qquad (8.18)$$

It was shown that regardless of the delay model used, the linear relations in the arrival time constraints lead to a following flow-like requirement on some of the Lagrangian multipliers:

$$\sum_{i \ \in \ \text{all outputs of gate } k} \lambda_{ki} = \sum_{j \ \in \ \text{all inputs of gate } k} \lambda_{jk} \qquad\qquad (8.19)$$

This leads to the restatement of the problem as

$$\text{minimize} \qquad L_\mu(\mathbf{x}) \qquad\qquad (8.20)$$
$$\text{subject to} \quad L_i \leq x_i \leq U_i \ \ \forall \ \text{gates} \ i$$

where

$$L_\mu(\mathbf{x}) = \sum_{i=1}^{n} \alpha_i x_i + \sum_{\text{all gates } i} \mu_i d_i \qquad (8.21)$$

$$\mu_i = \sum_{j \in \text{ inputs of } i} \lambda_{ji} \qquad (8.22)$$

This problem is then solved using Lagrangian relaxation. In each step, this iteratively sizes the gates in the circuit for the value of μ_i's from the previous iteration, exploiting the properties of the Elmore delay function to speed up the procedure. The initial values of the μ_i's are chosen to be uniform, and these values are adapted from step to step based on a subgradient optimization.

8.5.3 Sizing based on accurate nonconvex delay functions

The use of Elmore delay metrics has the disadavantage of low accuracy, while providing substantial advantages in terms of the convexity of the delay function. An alternative approach might use high accuracy delay evaluators that can guarantee that the solution can indeed achieve the performance promised by the optimizer. The drawback of such an approach, however, is the loss of the convexity properties of the Elmore metric. As a result, the result may fall in a local minimum rather than being guaranteed of a globally optimal solution.

The approach in [CEW+99, VC99] uses a fast and accurate simulator based on SPECS [VR91, VW93], which guarantees correctness of the delay computation. However, the simulator cannot guarantee the convexity of the delay metric, and therefore any such optimization comes at the cost of a potential loss in global optimality. To overcome this, the method uses a theoretically rigorous and computationally efficient optimization engine that is empirically seen to be likely to yield a good solution to the optimization problem.

The problem statement is similar to that in Equation (8.3), except that SPECS is used for delay computation to ensure better accuracy than an Elmore-based method. The use of SPECS removes the restriction that the optimization formulation must maintain convexity, and frees up the optimizer to incorporate several other constraints such as

- Upper and lower bounds on the signal transition times at each gate output

- Timing constraints related to dynamic logic

- Noise constraints may be incorporated: for example, sizing a gate very asymmetrically results in poor noise margins, and this may be stated as a constraint.

The LANCELOT optimization engine [CGT92] is used to perform the optimization, and an industrial strength implementation of the method is widely used in IBM. The engine requires evaluations of the objective and constraint functions and their gradients. The objective function, as in other methods, is simply the sum (or a weighted sum) of the transistor sizes, while most of the constraints are related to the results of a fast circuit simulation. For computational efficiency, the adjoint method [DR69] is used to calculate the gradients of constraint functions such as delay, slew, power, and noise.

The JiffyTune tool [CCH+98] differs from the methods listed so far, which rely on a static timing-based formulation, in that it employs detailed circuit simulations over a user-specified set of input patterns. The essential advantage of the method lies in its extreme accuracy; however, it places the burden of providing the worst-case input signals on the designer, and is appropriate at later stage of design, and for highly hand-crafted custom blocks. As in the work above[4] the simulation is driven by the SPECS simulator, and the optimization by the LANCELOT engine, which requires the evaluation of the objective and constraint functions, and their sensitivities. Sensitivity computations are performed using the adjoint network method, and one of the key observations that makes it efficient is in the adjoint Lagrangian method, which permits the sensitivity of a linear combination of functions to be computed by exciting the adjoint circuit with a train of impulses.

8.6   TIMING BUDGET BASED OPTIMIZATION

Several techniques (such as iCOACH [CK91], MOSIZ [DA89], and CATS [HF91]) are based on reducing the computation in the problem by heuristically solving it in two steps. The first step allocates a timing budget to each gate, while the second translates this timing budget into a set of transistor sizes. These two steps are repeated iteratively until the solution converges.

While all of the techniques cited above are purely heuristic, the method in [SSP02] provides guarantees of convergence under this framework when the delay function admits a simple monotonic decomposition (SMD) [SSP02]. The basis of this approach is the observation that for delay expressions with an SMD, such as the Elmore model, the gate delays and sizes are related by the expression

$$(D - A)\mathbf{x} = \mathbf{b} \tag{8.23}$$

where $\mathbf{x}$ is a vector of transistor sizes, A is a constant $n \times n$ matrix and $\mathbf{b}$ is a constant n-vector. The matrix D is an n-dimensional diagonal matrix whose diagonal elements correspond to the delay unknowns.

In the kth iteration, a small perturbation in $\mathbf{x}_k$ to $\mathbf{x}_k + \delta\mathbf{x}$ leads to a change in the delays, so that matrix D_k changes to $D_k + \delta D$. This change must obey the relation

$$(D_k + \delta D - A)(\mathbf{x}_k + \delta\mathbf{x}) = \mathbf{b} \tag{8.24}$$

From Equations (8.23) and (eq:vijay2), we obtain the following by neglecting the small $\delta D \delta\mathbf{x}$ term:

$$(D_k - A)(\delta\mathbf{x}) = \delta D\mathbf{x}_k \tag{8.25}$$

The two phases are described in greater detail below:

The D-phase assumes transistor sizes to be fixed and sets gate delay budgets ("D"). The objective function, which sums up all elements of $\mathbf{x}$, can be rewritten using Equation (8.25) to depend purely on the D variables. In addition, topological information must be incorporated: specifically, the slacks in the network are organized into a graph in which slacks may be allocated to gates. This results in a network flow formulation that can be solved to find the delay budgets.

The W-phase computes the optimal transistor sizes ("W") corresponding to these delay budgets by formulating the problem as a simple monotonic program [Pap98].

The CPU times provided by this method are, at the time of writing this book, the fastest demonstrated results for the transistor sizing problem.

8.7 GENERALIZED POSYNOMIAL DELAY MODELS FOR TRANSISTOR SIZING

8.7.1 Generalized posynomials

A major problem with the use of Elmore-based delay models is the limited accuracy that can be achieved by these models. Particularly in deep-submicron technologies and later, Elmore models leave much to be desired in terms of accuracy, although their convexity properties are well-beloved. As a result, the transistor sizing techniques described in Section 8.5.3 willingly abandoned the convexity properties of the Elmore model in favor of accurate, but nonconvex models. The idea of generalized posynomials for delay modeling was proposed in [KKS00] in an effort to provide accurate convex gate delay models for transistor sizing. The philosophy behind the approach is that posynomials and convex functions are a rich class of functions, and as described in Section 8.3.2, the posynomials that are used in Elmore-based delay models constitute a very restricted set where the exponents for each term of the posynomial belong to the set $\{-1, 0, 1\}$. Therefore, much of the space of posynomials and convex functions remains unexploited when Elmore models are used.

A generalized posynomial is defined recursively in [KKS00] as follows.

- A generalized posynomial of order 0 is simply a posynomial, as defined earlier in Equation (8.12).

$$G_0(\mathbf{x}) = \sum_j \gamma_j \prod_{i=1}^{n} x_i^{\alpha_{ij}} \qquad (8.26)$$

- A generalized posynomial of order $k > 0$ is defined as

$$G_k(\mathbf{x}) = \sum_j \gamma_j \prod_{i=1}^{m} [G_{k-1,i}(\mathbf{x})]^{\beta_{ij}} \qquad (8.27)$$

where $G_{k-1,i}(\mathbf{x})$ is a generalized posynomial of order less than or equal to $k - 1$, each $\beta_{ij} \in \mathbf{R}$, and each $\gamma_i > 0 \in \mathbf{R}$.

Like posynomials, generalized posynomials can be transformed to convex functions under the transform $(x_i) = (e^{z_i})$, but with the additional restriction that $\beta_{ij} \geq 0$ for each i, j.

For a generalized posynomial of order 1, observe that the term in the innermost bracket, $G_{0,i}(\mathbf{x})$ represents a posynomial function. Therefore, a generalized posynomial of order 1 is similar to a posynomial, except that the place of the x_i variables in Equation (8.12) is taken by posynomial. In general, a generalized posynomial of order k is similar to a posynomial, but x_i in the posynomial equation is replaced by a generalized posynomial of a lower order. **Example**: Consider a posynomial function

$$f_1(x_1, x_2, x_3) = 5x_1^{0.3}x_2^{-1.732} + 2.4x_2^2 x_3^{-1}$$

By definition, his is a generalized posynomial of order zero. An example of a generalized posynomial of order 1 is the function

$$
\begin{aligned}
f_2(x_1, x_2, x_3) = \ & 9.8 \cdot (7.6x_1^{7.6} + 5.4x_2^{5.4}x_3^{-5.4} + 3.2)^{1.09} + \\
& 2.7 \cdot (1.1x_1 + 2.2x_2 + 3.3x_3)^{5.5} \cdot (6.6x_1 + 7.7x_3 + 8.8)^{9.9} + \\
& 34.5 \cdot (1.2x_1x_2^{-3.3} + 3.4x_1x_2x_3 + 5.6x_3 + 7.8)^{1.732}
\end{aligned}
$$

A generalized posynomial that is not convex is

$$f_3(x_1, x_2) = \left(x_1^2 + x_2^2\right)^{-1} \tag{8.28}$$

This can be inferred from the fact that $f_3(x_1, x_2) \leq 1$ corresponds to the exterior of the unit circle, a nonconvex set; for a convex function, it must be true that such a set is convex [Lue84]. Note that the exponent of "-1" here does not satisfy the nonnegativity requirement that was described earlier.

For delay modeling, a curve-fitting approach is used to find a least-squares fit from the delay function, computed by SPICE over a grid, to a generalized posynomial in order to provide guarantees on accuracy of the delay model while using functions that are well-behaved in an optimization context. Under this framework, it is possible to incorporate constraints that take into account limitations on the allowable ratios between the n-block and the p-block transistors, which have repercussions on the noise margins.

8.7.2 Delay modeling using generalized posynomials

Outline of the delay modeling approach. Many commonly utilized delay estimation approaches, such as those used for standard cell characterization, estimate the delay of a gate for a given input transition time and output load capacitance, with the sizes of the transistors inside the gate being kept constant. However, for our purposes, any sizing procedure requires the timing model to capture the effects of varying the transistor sizes on the gate delays. This causes the number of parameters for the delay function to increase, making the problem of delay modeling for sizing algorithms more complex. These input parameters will be referred to as characterization variables.

We begin with an explanation of the timing model for an inverter, such as the one shown in Figure 8.5; this model is generalized to complex gates in subsequent sections. The aim is to be able to estimate delay as a function of the pmos and nmos transistor widths, w_p and w_n, the input transition time τ, and the output load capacitance, C_L. Therefore, for an inverter, w_p, w_n, τ, and C_L form the set of characterization variables. These variables reflect the set of variables that are generally considered to be important in defining the delay of a gate in most models.

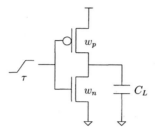

Figure 8.5. An inverter that is characterized using generalized posynomial delay models.

We attempted the use of several types of functions to achieve the desired levels of accuracy. The general form of expression that provided consistently good results for different gate types is as follows

$$\text{Delay} = \sum_{j=1}^{m} P_j \cdot \prod_{i=1}^{n} (x_i^{\Delta} + c_{ij})^{\beta_{ij}} + C \qquad (8.29)$$

Here, the x_i's are characterization variables, and the c_{ij}'s, β_{ij}'s, C, and P_j's are real constants, referred to collectively as *characterization constants*. The parameter Δ is set to either -1 or 1, depending on the variable, as will soon be explained. The problem of characterization is that of determining appropriate values for the characterization constants. We will show in Appendix C that the use of this form of function implies that the circuit delay can be expressed as a generalized posynomial function of the transistor widths.

Due to the curve-fitting nature of the characterization procedure (akin to standard cell characterization), it is not possible to ascribe direct physical meanings to each of these terms. However, it can be seen that the fall delay increases as C_L, w_p and τ are increased, and decreases as w_n is increased, implying that the value of Δ for the first three variables must be 1, and that for w_n should be -1. Note that this is not as restrictive as the Elmore form since, among other things, the β_{ij}'s provide an additional degree of freedom that was not available for the Elmore delay form. A similar argument may be made for the rise delay case.

Circuit simulations and curve-fitting.   A two-step methodology is adopted to complete the characterization. In the first step, a number of circuit simulations are performed to generate points on a grid. In the second, a least-squares procedure is used to fit the data to a function of the type in Equation (8.29).

A series of simulations is performed to collect the experimental data using a circuit simulator. The total number of data points, N, increases exponentially with the number of characterization variables. For the inverter circuit with four characterization variables and d data points for each variable to cover the range of interest, the total number of data points, N would be d^4. Therefore, it is important to choose the data points carefully; in particular, it is not necessary to choose an even grid for the transistor widths and a smaller granularity of points can be chosen for larger w_n's in case of the fall transition.

The determination of the characterization constants was performed by solving the following nonlinear program that minimizes the sum of the squares of the percentage errors over all data points.

$$\text{minimize} \sum_{i=0}^{N} a_i \left[\frac{D_{estim}(i) - D_{actual}(i)}{D_{actual}(i)} \right]^2 \qquad (8.30)$$

where N is the number of data points, $D_{estim}(i)$ and $D_{actual}(i)$, respectively, represent the values given by Equation (8.29), and the corresponding measured value at the ith data point. The parameter a_i is a user-settable parameter that is usually determined by running the simulation once with $a_i = 1 \forall i$, and then feeding back higher a_i values for points with higher errors. This nonlinear programming problem is solved using the MINOS optimization package [Dep95] to determine the values of characterization constants.

Characterization of a set of primitives. To illustrate the problem of directly extending the above methodology from inverters to arbitrary gates, we consider a three-input NAND gate circuit. The characterization variables for this gate will be the sizes of the three pmos transistors, sizes of the three nmos transistors, τ and C_L. If five data points were chosen to cover each of these variables, we would have total $5^8 = 32768$ data points. It is computationally expensive to perform such a large number of simulations and generate this database for curve fitting. For more complex gates, as the number of data points increases exponentially with the number of transistors in the gate, this would lead to a large overhead, both in terms of the simulation time and the time required to perform the curve-fit. Therefore, an alternative strategy is suggested. We emphasize even under this procedure, the transistor sizing approach will size each transistor individually, and this method is only only used for delay estimation.

It is important to emphasize that the use of these mapping strategies only serves to reduce the complexity of the characterization procedure. If one is willing to invest the CPU time required to perform the characterizations for each gate type, then this procedure is unnecessary. Since this is performed only once for each technology, it is viable to characterize all gate types that are expected to be used.

We now present a procedure that permits us to precharacterize a set of primitive logic structures accurately. The gate whose delay is to be measured is mapped to the closest precharacterized logic structure and the delay is calculated. A secondary advantage of this approach is that the delay model is not restricted to a fixed library type, and arbitrary gate types can be handled.

One straightforward technique that may be used is to map all of the gates to an "equivalent inverter" [WE93], and use the inverter characterization to estimate delays; the sizes of the pull-down nmos transistor and the pull-up pmos transistor of this inverter reflect the real pull-down or pull-up path in the gate. The widths of these new transistors are referred to as the equivalent widths. The equivalent width calculation is based on modeling the "on" transistors as conductances, and the equivalent width corresponds to the effective conductance of the original structure. Accordingly, if two transistors of widths w_1 and w_2 are connected in parallel, the equivalent width is defined as $w_1 + w_2$ and if the transistors are connected in series, the equivalent width is defined as $\left[w_1^{-1} + w_2^{-1} \right]$.

Two-input gates. The set of primitives used to approximate two-input gates are shown in Figure 8.6 (the presence of a load capacitance at the output is implicit and is not shown). The gates of the transistors are tied to either logic '0', logic '1', or to a switching input. This timing analysis procedure assumes only single input transitions, and hence there can only be one pair of pmos and nmos transistors switching at a time.

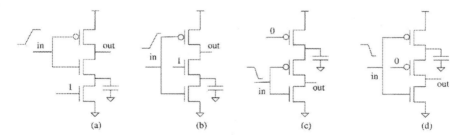

(a)   (b)   (c)   (d)

Figure 8.6.   A set of characterization primitives for two-input gates [KKS00].

Consider the two-input nand gate shown in Figure 8.7(a). For the fall delay, if the input transition occurs at input A, then the gate is mapped to Figure 8.6(a). Note that since the output is being pulled down in the case of a fall delay calculation, the pull-down is retained while pull-up is replaced by a single transistor, and the characterization equations of Figure 8.6(a) are used to estimate the delay. In a similar fashion, when the input transition occurs at input B of Figure 8.7(a), the gate is mapped to Figure 8.6(b). A similar procedure is applied for rise delays, i.e., the pull-up part is retained while the pull-down part is replaced by an equivalent nmos transistor. If we assume single input transitions, only one of the pmos transistors will be on during the rise output transition. The contribution of the pmos transistor that is off is neglected, and hence for rise delay calculation, the nand gate is mapped to an inverter. In a similar fashion, for the nor gate in Figure 8.7(b), when the transition is at input A, the gate is mapped to the primitive in Figure 8.6(d), and when the transition is at input B, the gate is mapped to the primitive in Figure 8.6(c).

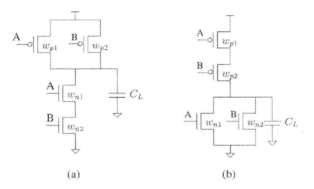

(a)   (b)

Figure 8.7.   Mapping two-input NAND and NOR gates to the primitives in Figure 8.6 [KKS00].

Complex gates.   For more complex gates, an expanded set of primitives is necessary. The set of primitives used to approximate complex CMOS gates is shown in Figure 8.8.

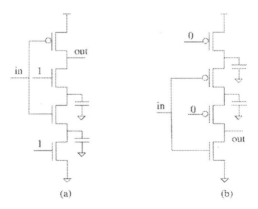

(a)   (b)

Figure 8.8.   Primitives for mapping complex gates [KKS00].

Before explaining the procedure of delay modeling, we introduce the notion of the *largest resistive path (LRP)*. In the worst case switching scenario for a gate, there is exactly one path from the output node to the ground [V_{dd}] node for a fall [rise] transition. This path may be formed by calculating equivalent widths for the "largest resistive path" from different nodes to ground/V_{dd} nodes [SRVK93]. The complex gate is represented by a directed graph with an edge from the drain and source nodes of each transistor in the gate[5]. Since the "on" resistance of a transistor is, crudely speaking, proportional to the reciprocal of its width, the edge weights are the reciprocals of the widths of corresponding transistors. The largest resistive path between nodes n_1 and n_2 is the path of largest weight from n_1 to n_2. The LRP and the weight of the LRP which corresponds to the width of equivalent transistor is found using a longest path algorithm [CLR90]. Note that the LRP computation based on the crude estimations of resistance are only required to predict the identity of the worst-case path, and more accurate delay modeling is carried out for the actual delay computation.

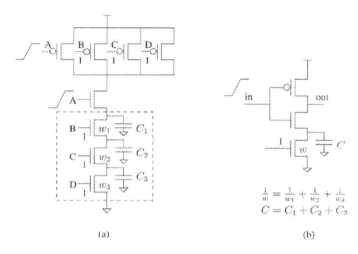

(a)   (b)

Figure 8.9.   An illustration of the procedure of mapping gates to primitives [KKS00].

We will now explain the computation of fall and rise delays for the gate.

FALL DELAYS: For fall delay estimation, the switching nmos transistor in the complex gate is identified; denote this transistor by N_{switch}.

Case A: If the source/drain node of the transistor N_{switch} is connected to the output node, the LRP from drain/source of N_{switch} to ground is replaced by its equivalent transistor, and the pull-up network by its equivalent transistor. The gate is thus mapped to a primitive as shown in Figure 8.9; this corresponds to a mapping to the primitive in Figure 8.6(a).

Case B: If the source/drain node of the transistor N_{switch} is connected to ground, the LRP from drain/source of N_{switch} to the output node is replaced by its equivalent transistor, and the pull-up network by its equivalent transistor. The gate is thus mapped to the primitive shown in Figure 8.6(b).

Case C: If neither the drain nor the source nodes of N_{switch} is connected to the V_{dd} or ground node, then the LRP 's from drain/source of N_{switch} to V_{dd}/ground nodes are replaced by their corresponding equivalent transistors.

The pull-up network is replaced by its equivalent transistor, thereby mapping the gate to the primitive shown in Figure 8.8(a).

RISE DELAYS: For rise delay estimation, the switching pmos transistor in the complex gate is identified. Let it be identified by P_{switch}.

Case A: If the source/drain node of the transistor P_{switch} is connected to the output node, the LRP from drain/source of P_{switch} to V_{dd} is replaced by its equivalent transistor, and the pull-down network by its equivalent transistor. The gate is thus mapped to the primitive shown in Figure 8.6(c).

Case B: If the source/drain node of the transistor P_{switch} is connected to V_{dd}, the LRP from drain/source of P_{switch} to output node is replaced by its equivalent transistor, and the pull-down network by its equivalent transistor. The gate is thus mapped to the primitive shown in Figure 8.6(d).

Case C: If neither the drain nor the source nodes of P_{switch} are connected to the V_{dd} or ground node, then the LRP 's from drain/source of P_{switch} to V_{dd}/ground nodes are replaced by their corresponding equivalent transistors. The pull-down network is replaced by its equivalent transistor, thereby mapping the gate to the primitive shown in Figure 8.8(b).

8.7.3 Statement of the problem

Ratioing constraints. Practical designs require that the rise and fall delays of a gate be balanced. However, this is not guaranteed by sizing algorithms, which may lead to unbalanced rise and fall delays. By limiting the ratio of pull-down to pull-up strength, the gate delays can be balanced. This idea is critical to the noise margins: for example, simple examples for the noise margin for an inverter (see Section 2.3 of [WE93]) show that the ratio of the pullup to the pulldown size must be controlled. A similar idea holds for more complex gates. The ratioing constraints can be mathematically expressed as

$$K_1 \leq \frac{\text{Pull-down strength}}{\text{Pull-up strength}} \leq K_2$$

Equivalently, this can be expressed as

$$\begin{aligned} \max \left(\frac{\text{Pull-down strength}}{\text{Pull-up strength}} \right) &\leq K_2 \\ \max \left(\frac{\text{Pull-up strength}}{\text{Pull-down strength}} \right) &\leq \frac{1}{K_1} \end{aligned} \qquad (8.31)$$

The sizing problem defined in (8.1) is reformulated with these constraints for every gate in addition to the original delay constraint. The term in the denominator is taken as the pull-down/pull-up strength of the equivalent inverter with the largest resistive path activated, and the term in the numerator is approximated as the equivalent size of the largest set of transistors that can be on in parallel in the opposite (pull-up/pull-down) network. The terms K_1 and K_2 are user-settable constants.

The optimization problem. The problem that we solve is an altered form of the traditionally stated problem of (8.1). In addition to minimizing the cost function under delay constraints at primary output nodes, constraints are placed on the transition delay for the transitions at internal nodes in the circuit and ratioing constraints. The addition of constraints on the transition delay, apart from improving the delay properties of the circuit, also serves to control the short-circuit power dissipation. Therefore, the power dissipated by the circuit may be measured in terms of the dynamic power, and short-circuit power effects described in [SC95] need not be considered. The problem may be stated succinctly as follows:

$$\text{minimize} \qquad Area \text{ or } Power \tag{8.32}$$

$$\text{subject to} \qquad Delay \leq T_{spec} \text{ at all primary outputs}$$

$$\tau_i \leq \tau_{spec} \ \forall \ \text{gate outputs } i$$

$$K_1 \leq \frac{\text{Pull-down strength}}{\text{Pull-up strength}} \leq K_2 \ \forall \ \text{gates } i$$

A proof of convexity of this formulation is provided in Appendix C. Although this is somewhat theoretical, it is recommended that the reader peruse this, since the proof also points to several practical methods that could be used to preserve the convexity of the formulation.

In priciple, the accuracy of the model may be increased by enhancing the richness of the modeling functions. For the transition time expression, a sum of terms could be used and this could be shown to maintain the convexity properties of the formulation. For delay modeling, one could use the "max" operator to further enhance accuracy, if required. If $f_1, f_2, \cdots, f_k$ are convex functions then the function $F = \max(f_1, f_2, \cdots, f_k)$ is also a convex function and hence can be used to improve accuracy of the approximation, while retaining the convexity properties of the delay function. However, to implement such a scheme, the number of characterization constants will increase, thereby increasing the characterization effort. Nevertheless, it may be a useful extension if the accuracy achieved using current model is not deemed sufficient. Since the current modeling approach is a special case of this, where $k = 1$, the extension is guaranteed to do no worse than this approach.

8.8   DUAL V_T OPTIMIZATION

The use of dual threshold voltage (V_t) values presents another approach for optimizing the delay of a circuit. Unlike transistor sizing, however, it presents no significant area overhead; however, its use has ramifications on the power dissipation of a circuit, specifically the leakage power. A transistor is an imperfect switch, and even in the off state, some current can leak through it.

In recent technologies, the power dissipation associated with the leakage current has grown increasingly important, and in particular, when a lower V_t is used for a device, it makes the switch even more imperfect and increases the leakage power[6]. On the other hand, using a lower V_t for a device increases its current drive and allows for faster switching speeds and hence reduced delays. Therefore, if dual V_t values are used, it is possible to trade off the increase in the leakage power with the delay reduction.

Various methods for V_t optimization have been presented, such as those in [WCJ+98, WV98, WCR+99, SP99], all of which take a circuit and alter transistor V_t values for a leakage-delay tradeoff. Alternative methods that look at the problem of concurrently sizing the transistors in a circuit and assigning V_t values are presented in [SEO+02, KS02].

The leakage of a gate depends on the logic signals at its inputs, and different input combinations ("states") correspond to different leakage values. The work in [SEO+02] defines the concept of a *dominant leakage stage*: the largest leakage values are seen when only one transistor in a series chain is off (i.e., its gate is at logic 1 if it is a pmos device, or at logic 0 if it is an nmos device). Therefore, these states are referred to as dominant states. When more than one transistor in a series chain if off, the corresponding leakage for such a state is significantly smaller than the leakage in a dominant state. This observation considerably simplifies the leakage computations.

The work in [SEO+02] then proceeds to present a heuristic for simultaneous V_t and sizing optimization. The problem is formulated as one of delay-leakage tradeoffs at a constant area. The extreme points for the delay correspond to the all-low and all-high V_t assignments. The procedure consists of two steps: in the first, a merit function is evaluated to determine the largest delay reduction for the smallest leakage increase. The transistor with the best merit function value is then set to low V_t. The second step is one of rebalancing: with the change in the V_t, the previous device sizes are longer optimal since alterations in V_t can change both the speed and the capacitance of the gate. Therefore, a resizing is carried out: this first reduces the area due to oversized devices that contribute to large slacks, and then redistributes this recovered area to achieve a reduction in the maximum delay of the circuit, performing delay reduction until the original area is reached.

An alternative approach in [KS02] examines the utility of sizing and V_t assignment in terms of the leakage-delay tradeoff, without the constraint of keeping the area constant. The algorithm is based on the following observations:

- A typical delay-area curve for a circuit has the largest delay when all transistors are at minimum size. As the transistor sizes are altered, the delay first increases rapidly for small area increases. This continues up to a "knee point," beyond which further delay reductions entail massive area increases.

- The delay of a gate decreases linearly with increasing width, but superquadratically with decreasing V_t. This analysis does not count the effects of the increased width on the delay of the previous stage of logic; this will reduce the trend to less than linear for sizing. Therefore, it can be seen that V_t reduction is more efficient at reducing the circuit delay than sizing.

- The leakage current varies exponentially with V_t, but linearly with the transistor size. This implies that as long as delay reductions can be achieved by small increases in transistor areas, sizing is preferable to lowering V_t.

Putting these together, and starting with a circuit with all transistors at minimum size and high V_t, one may conclude that up to the knee point, where the transistor area increase is small, delay reduction should be achieved through sizing. Beyond this point, the leakage increase due to both sizing and V_t reduction increases in an exponential-like manner, so that it is preferable to use V_t reduction, which can deliver a larger delay reduction. The procedure uses the TILOS heuristic to reach the knee point, and then devises an enumeration strategy with aggressive pruning to perform V_t assignments.

8.9   RESOLVING SHORT PATH VIOLATIONS

Correct circuit operation requires that all data paths have a delay that lies between an upper bound and a lower bound to satisfy long path and short path constraints. Traditional approaches in delay optimization for combinational circuits, such as those reviewed in the rest of this chapter, have dealt with methods to decrease the delay of the longest path, frequently neglecting to consider the problem of short path violations that may already exist, or may be introduced as a result of the optimizations.

In [SBS93], the issue of satisfying the lower bound delay constraints for edge-triggered circuits is addressed. The short path constraints are handled as a post-processing step after traditional delay optimization techniques by padding delays into the circuit to ensure that it meets all short path constraints. The problem of inserting a minimum number of such delays is referred to as the minimum padding problem. Two methods for solving the problem are described in [SBS93]:

(1) The first simple algorithm finds an edge on a short path that does not lie on any critical long path. A delay is inserted on the edge until one of the following two cases becomes true:

    (a) either some path containing the edge becomes a critical long path, in which case no further delays are added on this edge, and the process is continued for the edge that is identified next, or

    (b) the short path containing the edge meets the lower bound and thus success is achieved in padding.

This procedure is carried for all the short paths until the lower bound delay constraints are met. This procedure is purely a heuristic and has no guarantees of optimality.

(2) The second approach formulates the problem as a linear program. The combinational circuit is represented as a directed acyclic graph with every gate output, primary input or primary output in the circuit represented by a vertex in the graph. Two type of edges are defined, namely, internal and external edges, denoted by E^I and E^E. An internal edge is directed from an input pin of a gate to its output pin. The external edges represent the interconnections or wiring in the circuit. Hence, if an edge $e_{i,j} \in E^I$, then the weight $w_{i,j}$ of the edge is the delay of the gate from the input represented by i to the output of the gate. If the edge $e_{i,j} \in E^E$, then the weight $w_{i,j}$ of

the edge is a variable whose value is to be determined; the physical meaning of the variable $w_{i,j}$ here is the number of delays padded on the edge $e_{i,j}$. Delays may be inserted only on the external edges. There are two arrival times associated with each vertex i: an early arrival time a_i and a late arrival time A_i. If i is a primary input (PI), then a_i and A_i are specified by the user or the algorithm at the higher level, say, as $a_i = \lambda_i, A_i = \Lambda_i$. If i is an input/output pin of a gate, then

$$a_i = \min_{j \in FI(i)} (a_j + w_{j,i})$$
$$A_i = \max_{j \in FI(i)} (A_j + w_{j,i}) \qquad (8.33)$$

where $FI(i)$ denotes the set of fanins of i. A path is a sequence of vertices, and the delay of a path p is denoted by $d(p)$. At every primary output (PO) i, the data is required to be available no earlier than r_i and no later than R_i; namely, $r_i \leq a_i \leq A_i \leq R_i$. A path p fails to meet the short path constraint if $a_i + d(p) \leq r_i$. In that case, delays $w_{i,j}$ are added along the edge (i, j) on the short path to remove this violation. The area overhead of this addition is proportional to the number of delays added, and is approximated as $\sum_{e_{i,j} \in E^E} w_{i,j}$. The problem of finding the minimum-cost padding that satisfies the timing constraints can now be expressed as a linear program, with constraints obtained by relaxing Equation (8.33), and an objective function that corresponds to the number of buffers added.

$$\text{minimize} \qquad \sum_{e_{i,j} \in E^E} w_{i,j}$$

$$\text{subject to} \quad A_j \geq A_i + w_{i,j} \quad \forall i \in FI(j)$$
$$a_j \leq a_i + w_{i,j} \quad \forall i \in FI(j)$$
$$A_i \leq R_i \quad \forall i \in PO$$
$$a_i \geq r_i \quad \forall i \in PO$$
$$A_i = \Lambda_i \quad \forall i \in PI$$
$$a_i = \lambda_i \quad \forall i \in PI \qquad (8.34)$$

It was shown that in any circuit, if for every pair $i_1 \in PI$ and $i_n \in PO$ such that $\exists$ a combinational path $p : i_1 \rightsquigarrow i_n$, we are given that $A_{i_1} - a_{i_1} \leq R_{i_n} - r_{i_n}$, then the padding problem can be solved. An intuitive explanation for this condition is as follows. If we interpret the term, $A_{i_1} - a_{i_1}$, as the uncertainty interval in the arrival time of the signal at the primary input, i_1, and the term, $R_{i_n} - r_{i_n}$, as the uncertainty in the required time at the output, then since the circuit is causal, it cannot make the uncertainty interval at the output any narrower than the uncertainty interval at its input.

8.10   SUMMARY

This chapter has overviewed several methods that can be used for transistor-level timing optimization of digital circuits. The specific methods overviewed include transistor sizing, dual V_t optimization, and padding for short paths. These are intended to provide a flavor for timing optimization and should not be construed as a complete review of such methods. In addition to these, numerous methods for interconnect optimization using buffer insertion and wire sizing have been proposed in the literature. Even more such techniques are on the way: for instance, recent technologies have seen large increases in the gate leakage (as opposed to the subthreshold leakage, addressed in Section 8.8), which can be translated into a leakage/delay tradeoff problem by using dual oxide thickness (T_{ox}) values.

Notes

1. The current drive of a gate is controlled by the ratio of the channel width to channel length, and if channel lengths are not uniform, this ratio can be considered as the size. However, this may result in more complex functional forms for the area or power during optimization.

2. This was subsequently extended to an RC tree model in iCONTRAST [SRVK93] to model capacitors off the largest resistive path that must be charged/discharged.

3. In practice, when the effects of input transition times are considered under a more accurate model, one may also have to consider a few stages downstream of the gate in which the transistor lies.

4. It should be noted that Jiffytune predates [CEW+99, VC99].

5. In a static CMOS gate, it is always possible to uniquely identify the source node and the drain node. This may not be true in circuits with pass gates, which are not handled in this work. The Jouppi rules [Jou87a], for example, could be used to extend this work to circuits with pass gates.

6. This refers to the leakage mechanism referred to as *subthreshold leakage*, where the nonideal nature of the transistor switch implies that it conducts a nonzero current in the region where the gate-to-source voltage for an nmos transistor (or vice versa for a pmos transistor) is below V_t, in a regime where the transistor is supposedly off.

9 CLOCKING AND CLOCK SKEW OPTIMIZATION

9.1 ACCIDENTAL AND DELIBERATE CLOCK SKEW

Conventional synchronous circuit design is predicated on the assumption that each clock signal of the same phase arrives at each memory element at exactly the same time. In a sequential VLSI circuit, due to differences in interconnect delays on the clock distribution network, this simultaneity is difficult to achieve and clock signals do not arrive at all of the registers at the same time. This is referred to as a skew in the clock. In a single-phase edge-triggered circuit, in the case where there is no clock skew, the designer must ensure that for correct operation, each input-output path of a combinational subcircuit has a delay that is less than the clock period. In the presence of skew, however, the relation grows more complex and the task of designing the combinational subcircuits becomes more involved.

The conventional approach to design builds the clock distribution network so as to ensure zero clock skew. An alternative approach views clock skews as a manageable resource rather than a liability, and uses them to advantage by intentionally introducing skews to improve the performance of the circuit. To illustrate how this may be done, consider the circuit shown in Figure 9.1, and assume each of the inverters to have a unit delay. This circuit cannot be properly clocked to function at a period of 2 time units, because as shown in the figure, the required arrival time of the signal at register L1 is 2 units, while the data arrives after 3 time units. It is readily verifiable that the fastest allowable clock for this circuit has a period of 3 units. However, if a skew of +1 unit is applied to the clock line to register L1, the circuit can operate under a clock period of 2 units. This is possible because as shown in Figure 9.2, the application of a skew of +1 unit delays the clock arrival at register L1 by one unit, thus changing the required data arrival time to the new arrival time of the first clock tick, which is 3 units (i.e., the period of 2 unit delayed by +1 unit). Under these circumstances, the actual data arrival time of 3 units does not cause a timing violation, and the circuit is correctly clocked. A formal method for determining the minimum clock period and the optimal skews was first presented in the work by Fishburn [Fis90], where the clock skew optimization problem was formulated as a linear program that was solved to find the optimal clock period.

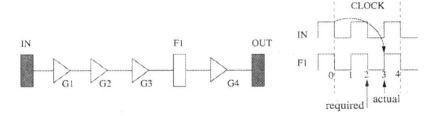

Figure 9.1.   An example edge-triggered circuit: if each gate has a unit delay and the setup time of each flip-flop is zero, then its optimal clock period is 3 units.

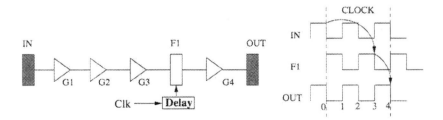

Figure 9.2.   The advantage of nonzero clock skew: by introducing a unit delay in the clock feeding the central flip-flop, a clock period of two units can be achieved.

A common misconception about changing clock skews is that it is believed to be an "unsafe" optimization, in that a small change in the gate/interconnect delays may cause a circuit with precariously small tolerances to malfunction. In fact, this is not so; one can build in safety margins, as shown in Section 9.3, that ensure that skewing errors do not disrupt circuit functionality. These margins ensure that the circuit will operate in the presence of unintentional process-dependent skew variations. In fact, introducing deliberate delays within the clocking network has been a tactic that has long been used by designers [Wag88] to squeeze extra performance from a chip, sometimes in a somewhat clandestine manner. Only in the last few years has this idea become more "mainstream," so that a designer may feel free to try it in full public view. From the design automation point of view, zero-skew clock tree construction techniques may be adapted to build fixed-skew clock networks. A proof of the practicality of this concept was demonstrated in a pipelined data buffer chip using the concept of skewed clocks, which was designed and fabricated in [HW96].

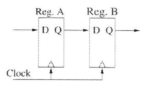

Figure 9.3. An example illustrating the perils of zero skew, when two flip-flops are connected without any combinational delay between them.

It is, in fact, a misconception to believe that zero skew is entirely safe. To see this, consider a shift register consisting of register A whose output is connected to register B with no combinational logic between the two [SSF95], as shown in Figure 9.3. Even for a circuit designed for zero skew, a small unintentional positive skew at register B will cause double-clocking, i.e., a short path constraint violation. Such problems may be avoided by the use of the safety margins mentioned in the previous paragraph and the introduction of deliberate nonzero skew: a small amount of deliberate positive skew at A provides an effective safety margin against short path violations.

The organization of this chapter is as follows. We first briefly describe techniques that may be used to build zero-skew and fixed-skew clock trees. This is followed by clock skew optimization algorithms for period optimization, keeping the structure of the remainder of the circuit unchanged. Finally, the use of clock skew optimization in conjunction with transistor sizing, where the parameters of the circuit are changed at the same time as the optimal skews, and its use for ground bounce reduction, are presented.

9.2 CLOCK NETWORK CONSTRUCTION

The clock distribution network is required to propagate the clock signal to every register in a synchronous circuit. Consequently, it must be designed carefully to ensure the optimal use of resources and to achieve performance requirements on the clock signal at the sink nodes. These performance requirements include requirements on the signal arrival time and on the sharpness of the clock edge. The skew constraint on clock networks is a hard constraint, and to ensure the tractability of the problem, many clock networks are constructed as trees.

In this section, we will present the essentials of clock tree construction methods targeted at achieving zero skew and fixed skew. Although they will not be discussed here, it is worth pointing out that meshes have been proposed for clock distribution, for example, in [DCJ96, XK95, SS01]. This section is only intended to be an abbreviated reference on clock tree construction; for details, the reader is referred to [KR95, Fri95, CHKM96]. As we will see, hybrid tree/mesh structures are also commonly used in high-performance design.

9.2.1   H-tree-like approaches

Early methods for clock tree design attempted to achieve zero-skew by balancing the wire length to each sink. H-trees [FK82], illustrated in Figure 9.4 and so named because of their structure, were recognized for years as a technique to help reduce the skew in synchronous systems. The H-tree ensured, by construction, that all sinks of the clock tree are equidistant from the clock source. For regular structures such as systolic arrays, the H-tree works well to reduce skew, but in the general case, asymmetric distributions of clock pins are common and a symmetric H-tree may not be effective for clock routing. To address this issue, the method of recursive geometric matching [KCR91] and the method of means and medians [JSK90] were proposed; these extend the H-tree idea to structures with a smaller degree of regularity.

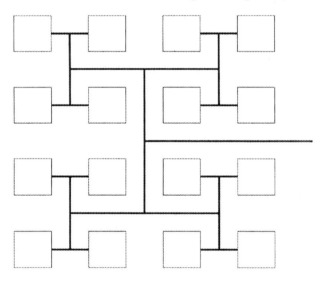

Figure 9.4.   An illustration of an H-tree layout for clock distribution.

Nevertheless, H-trees have still been used very widely because of their regular structure, which makes it easy to account for routing resource utilization. It is common in high-performance design to build a backbone structure that is based on an H-tree and to size the wires suitably so as to achieve zero skew [PMP93, PMOP93].

9.2.2   Exact zero-skew clock routing

Early H-tree based methods used symmetric structures, and simplified the problem of equalizing the clock delays into the geometric problem of equalizing the path lengths on a Manhatan grid. This assumption is valid for a perfectly symmetric clock network in which the load capacitances at all the sinks is precisely the same; however, this is an invalid assumption for real circuits. The work in [Tsa91, Tsa93] recognized that due to uneven loading and buffering effects, path-length equalization methods do not achieve the effect of balancing clock delays. An approach that incorporates this effect and uses the Elmore delay model to automatically build zero-skew clock trees was suggested. We will now describe this method in detail, and explain how it may be extended to build clock trees with specified, and possibly nonzero, skews at the sink nodes.

The Elmore delay model [RPH83] is used to calculate the signal propagation delay from a clock source to each clock sink. A modified hierarchical method is proposed for computing these delays in a bottom-up fashion. This algorithm takes advantage of the structure of the Elmore delay calculations to construct a recursive bottom-up procedure for interconnecting two zero-skew subtrees to form a new tree with zero skew.

We now describe one recursive step of the bottom-up procedure. In each such step, two subtrees are combined into one subtree; by construction, it is ensured that each such subtree has zero skew from its current root to all of its sinks. In other words, the signal delay from the root to the leaf nodes of the subtree are forced to be equal. In the first step, each sink node constitutes a subtree, and the skew within each such subtree is trivially seen to be equal to zero. In subsequent steps, partially constructed zero-skew subtrees are combined.

In one step of the recursion, two zero-skew subtrees are connected with a wire and a new root is defined for this combined structure, ensuring zero skew from the root to all sinks in the merged tree. This is illustrated in Figure 9.5, and this new root is referred to as the *tapping point*. If we denote the total wire length of this connecting wire segment as l, and the wire length from the tapping point to the root of subtree1 be $x \times l, 0 \leq x \leq 1$, then the wire length from the tapping point to the root of subtree2 is $(1 - x) \times l$.

A closed-form expression for the value of x that yields zero skew will now be determined. Consider each subtree in Figure 9.5. Since subtree1 [subtree2] has zero skew, the delay from its root to each leaf node is equal, say t_1 [t_2]. Let the corresponding total capacitances in the subtrees be C_1 [C_2], respectively. The wires of length $x \times l$ and $(1 - x) \times l$ are modeled using the RC π model shown in the figure.

The delay from the tapping point to each leaf node in subtree1 is equated to the delay to each node in subtree2 using the Elmore delay formula [SK93] to provide the following equation

$$r_1(\frac{c_1}{2} + C_1) + t_1 = r_2(\frac{c_2}{2} + C_2) + t_2 \qquad (9.1)$$

Substituting $r_1 = \alpha x l$, $r_2 = \alpha(1 - x)l$, $c_1 = \beta x l$, and $c_2 = \beta(1 - x)l$, where α and β are, respectively, the per unit resistance and capacitance of the intercon-

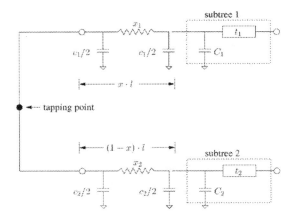

Figure 9.5.   Zero-skew merge of two subtrees in the algorithm of [Tsa91].

necting wire, we find that the tapping point corresponds to

$$x = \frac{(t_2 - t_1) + \alpha l \left(C_2 + \frac{\beta l}{2} \right)}{\alpha l (\beta l + C_1 + C_2)} \tag{9.2}$$

If the solution of Equation (9.2) results in a value x that satisfies $0 \le x \le 1$, then the tapping point lies along the segment interconnecting the two subtrees and is legal. In case x lies outside this range, it implies that the loads on the two subtrees are too disparate. If so, the tapping point is set to lie at the root of the subtree with the larger loading, and the wire connected to the other segment is elongated in order to ensure that its additional delay balances the delay of the first subtree. This procedure of elongating the wire is referred to as *snaking*. If $x < 0$, then the tapping point is kept at $x = 0$ and the connecting wire is elongated by an additional length l'. The value of l' is found by matching the delays in each subtree, and is given by

$$l' = \frac{\sqrt{(\alpha C_1)^2 + 2\alpha\beta(t_2 - t_1)} - \alpha C_1}{\alpha\beta} \tag{9.3}$$

The formula for l' for the $x > 1$ scenario can be derived in a similar manner.

Thus, in this approach the delay time to each node is derived from its immediate predecessor, the branch resistance, and the subtree capacitance. Buffering effects are accounted for in this procedure by storing C_1 and C_2, as the total subtree capacitance up to their closest downstream buffer locations; buffers are added to ensure that the clock edge remains sharp.

9.2.3   Further refinements

The procedure described by the algorithm described above does not attempt to reduce the length of the zero-skew clock tree. The deferred-merge embedding

(DME) algorithm [BK92, CHH92, Eda91] substantially reduces the tree cost while satisfying the skew specifications. In each step of the zero-skew merging procedure described earlier, the two subtrees are connected by a wire of length l. In a Manhattan geometry, there are several routes between two points that have the exact same length. Therefore, the precise route taken by the wire connecting the two zero-skew subtrees is important in that it affects the optimality (but not the correctness) of future merge operations, where the optimality is defined in terms of minimizing the wire length of the clock tree.

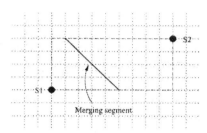

Figure 9.6.   A merging segment between two points, with $x = \frac{5}{13}$, in the construction of a zero-skew tree.

The DME procedure works by deferring the precise route, or embedding, corresponding to each node for as long as possible, and operates in two phases. In the first bottom-up phase, a tree of "merging segments" is constructed, where each merging segment is the locus of possible placements of a specific tapping point in the clock tree, under a Manhattan geometry. An example of a merging segment for two points, S1 and S2, is shown in Figure 9.6, with the value of x in Equation 9.2 set to $\frac{5}{13}$. It should be pointed out that the locus of points that form a merging segment for a pair of straight lines segments is another straight line segment; this implies that as the method goes up the recursive merging process, the locus of merging points is always given by a straight line. In the second top-down phase, the exact location of these internal nodes are resolved with the purpose of minimizing the total wire length of the clock tree.

Apart from reducing the delay by minimizing the wire lengths and inserting buffers, one may also use wire width optimization for sharpening the clock edge. Techniques for this purpose have been presented in, for example [Eda93, PMP93, PMOP93]. The problem of optimally inserting buffers in clock trees has been addressed in [CS93b, TS97, VMS95].

The zero-skew merge procedure is extended to RLC clocking networks in [LS96, LS98], using higher-order moment matching methods that are a generalization of the Elmore delay modeling approach. The procedure ensures adequate signal damping and satisfies constraints on the clock signal slew rate.

9.2.4   Fixed skew clock network construction

The procedures described above can be extended [Tsa93] to build a clock tree that is required to deliver the clock with a predetermined skew at each sink. This problem is termed the fixed-skew clock network construction, or the prescribed-skew routing problem. As described earlier in this chapter, the major utility of nonzero skews is to enable time-borrowing between combinational circuit segments that are connected by a flip-flop.

The technique of [Tsa91, Tsa93] (and its extensions) can be adapted to solve this problem by adding a fictitious delay element on each clock sink. For a prescribed skew of $+D_i$ at the ith element, the algorithm begins with each sink node corresponding to a (trivial) tree with a delay of $-D_i$, instead of zero as before. Note that according to the notation, a positive skew $+s$ implies that the clock signal is delayed by time s.

Therefore, in principle, in terms of the design automation effort, it is no harder to design fixed-skew clock trees than it is to build zero-skew trees. Several other papers on fixed-skew clock network construction have been published, including [HLFC97, NF96, NF97]. The work in [NF96, NF97] develops a technique that first determines the optimal skew schedule for a circuit, and then constructs a fixed skew network by using the delays of wires and by inserting buffers for additional delays, while minimizing sensitivity to process and environmental delay variations. The approach in [HLFC97] is distinguished by the fact that it performed a detailed circuit-level design and proved the viability of the process through a fabricated chip.

9.2.5   Clock networks in high performance processors

In this section, we will provide a brief description of clock networks that have been used in a few high performance processor families. While many of these are aimed at providing zero skew, some may be extended for nonzero skew clock network construction. This description will largely present the simplest overview. The reality may be have a few additional complications since many of these chips work in multiple clocking domains. For a clear example of how this is handled, the reader is referred to [KBD$^+$01]; several of the papers referenced below also provide further details on how multiple clocking domains are handled.

The DEC Alpha clock network.   The alpha processors were among the first to use meshes instead of trees, in an effort to control the clock skew. The philosophy behind this construction is that the sensitivity of the clock skew to process variations is reduced in a mesh that is driven redundantly by multiple drivers. For the case of the 600 MHz Alpha processor [BB98], the master clock signal from a phase-locked loop (PLL) is fed to a global clock grid through an X-tree or H-tree to 16 distributed global clock (GCLK) drivers. The lowest level of GCLK drivers are in the form of a "windowpane" arrangement, as shown in Figure 9.7(a), and the last two levels are optimized so that the delay from the PLL to the windowpanes is identical. The windowpanes drivers are connected to a dense global clock mesh, which is redundantly and multiply driven by the clock signal, as shown in Figure 9.7(b). Finally, the clocks are taken to the utilization points from this mesh. A design automation strategy for sizing wires in clock meshes was presented in [DCJ96], and was applied to an earlier generation processor. A new clocking procedure for the 1.2GHz Alpha processor is described in [XBG+01], and uses DLLs to compensate for process variations.

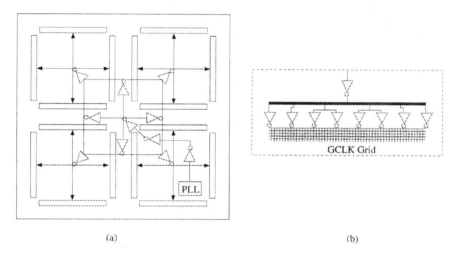

(a)   (b)

Figure 9.7.   The clock distribution network in the 600 MHzAlpha processor. The overall structure is shown in (a), and a detail of the windowpane driver in (b) [BB98].

The IBM Power4 clock network.   The strategy for the 1.3 GHz IBM Power4 also uses clock meshes at the lowest level, which are driven by buffered tunable trees or tree-like networks at upper levels [RMW+01, RCE+02]. The PLL drives a global clock distribution tree structure that delivers the clock signal from the PLL near the center of the chip, through an H-tree with four levels of buffers, to 64 sector buffers. Each sector buffer drives a final tree of tunable wires, which in turn, drive a single grid, which covers the entire chip, at 1024 points.

The Intel IA-64 clock network.   The clock network for the first IA-64 processor consists of three levels [TRD+00]: a global H-tree based distribution level where the PLL output is sent to a set of deskew buffers (DSKs), a regional distribution level where the DSK outputs are distributed to 30 regional clock grids (meshes) through regional clock drivers (RCDs), and a local distribution level where the local clock buffers driven by the regional clock grid provide the clock signal to clocked elements, including the possibility for providing skewed clocks[1]. This is pictorially shown in Figure 9.8(a). The design of the DSK is

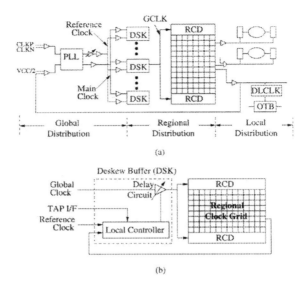

Figure 9.8. An illustration of the clock network in an IA-64 processor. The overall scheme is shown in (a), and the deskew buffer architecture in (b) [TRD+00].

particularly interesting: a separate reference clock is distributed along with the core clock distribution, and feedback clocks are provided from the end points of the core clock distribution back to the DSK, as shown in Figure 9.8(b). A phase detector in the DSK then samples the phase difference between the reference clock and the local feedback clock, and a tunable delay circuit is used to adjust the skew until it is removed. Early generations of the processor selected one of several discrete delays by picking the appropriate tap from a delay line, but more recently, fuse-based deskew has been employed.

9.3   CLOCK SKEW OPTIMIZATION

It has been demonstrated at the beginning of this chapter that the use of intentional nonzero clock skews can speed up the clock period for a sequential circuit since it facilitates cycle-borrowing. The work of Fishburn [Fis90] formalized an approach to find the optimal skews at each memory element in an edge-triggered circuit, where the optimality was defined in terms of minimizing the clock period. The problem was posed as a linear program, and this formulation is described in this section.

9.3.1   Timing constraints

Consider a combinational block in a sequential circuit, as shown in Figure 9.9. Let FF_i and FF_j be a pair of flip-flops at the input and output, respectively, of the combinational block, with skews of x_i and x_j, respectively.

Denoted the delay of the combinational block between them as $d(i,j)$, with $\underline{d}(i,j)$ being the minimum delay and $\overline{d}(i,j)$ being the maximum delay. If T_{hold} is the flip-flop hold time, T_{setup} the flip-flop setup time, and P the clock period, then in the presence of skews, the timing constraints take the following form

Figure 9.9.   A combinational block in a sequential circuit.

Long path constraints To avoid "zero-clocking," the data from the current clock cycle should arrive at FF_j no later than a time T_{setup} before the next clock. Since the data leaves FF_i at time x_i, the latest time by which it will reach FF_j is $x_i + \overline{d}(i,j)$. The long path constraint, shown in Figure 9.10, may be expressed as

$$x_i + \overline{d}(i,j) \leq x_j + P - T_{setup} \tag{9.4}$$

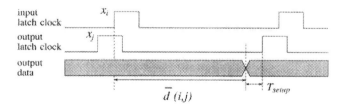

Figure 9.10.   A pictorial depiction of long path constraints.

Short path constraints To avoid "double-clocking," the data from the next clock cycle should arrive at FF_j no earlier than a time T_{hold} after the current clock. The data leaves FF_i at time x_i and, therefore, the earliest time it can reach FF_j is $x_i + \underline{d}(i,j)$. Thus, the short path constraint illustrated in Figure 9.11 is

$$x_i + \underline{d}(i,j) \geq x_j + T_{hold} \tag{9.5}$$

9.3.2   Clock period minimization

The problem of minimizing the clock period P by controlling the clock skew at each flip-flop, subject to correct timing, can now be formulated as the following

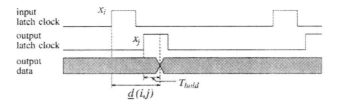

Figure 9.11.   An illustration of short path constraints.

linear program

$$\begin{aligned}
\text{minimize} \quad & P \\
\text{such that} \quad & x_i + \underline{d}(i,j) \geq x_j + T_{hold} \\
& x_i + \overline{d}(i,j) \leq x_j + P - T_{setup}
\end{aligned} \tag{9.6}$$

The linear program is formulated under the assumption that the delays of the combinational segments are constant. These delays can be calculated efficiently by using the method of Section 5.2.1.

The above formulation may be altered to model uncertainties in the clock skews. If it can be guaranteed that in the manufactured circuit, the skew at flip-flop k will be within the range $[x_k - \delta/2, x_k + \delta/2]$, where x_k is the designed value of the skew, then the difference between any skews, $(x_i - x_j)$ in the manufactured circuit, must be within δ of the designed value of $(x_i - x_j)$. A linear program that is guaranteed to produce a functional manufactured circuit is formulated below:

$$\begin{aligned}
\text{minimize} \quad & P \\
\text{such that} \quad & x_i + \underline{d}(i,j) \geq x_j + T_{hold} + \delta \\
& x_i + \overline{d}(i,j) \leq x_j + P - T_{setup} - \delta
\end{aligned} \tag{9.7}$$

An alternative formulation to maximize the tolerance of the solution to unpredictable changes, for a given clock period P, is as follows. This may be achieved by maximizing the minimum slack over all the constraints, converting the problem into a minmax problem. A new variable M is introduced, and is added to each of the main constraint inequalities, so that maximizing M is tantamount to finding the skew values that maxime the minimum slack over all the inequalities. The precise formulation is as shown below:

$$\begin{aligned}
\text{maximize} \quad & M \\
s.t. \quad & x_i + \underline{d}(i,j) - M \geq x_j + T_{hold}, \\
& x_i + \overline{d}(i,j) + T_{setup} + M \leq x_j + P
\end{aligned} \tag{9.8}$$

9.3.3 Efficient solution of the skew optimization problem

The constraints of the linear program (9.7) are rewritten as

$$x_i - x_j \geq T_{hold} - \underline{d}(i,j) + \delta$$
$$x_j - x_i \geq T_{setup} + \overline{d}(i,j) - P + \delta, \qquad (9.9)$$

Note that the skews at the primary inputs and the primary outputs may be set to zero under the assumption that they cannot be controlled or adjusted. For a constant value of P, the constraint matrix for this problem reduces to a system of difference constraints [CLR90]. For such a system, it is possible to construct a constraint graph. The vertices of the constraint graph correspond to the x_i variables, and each constraint $x_a - x_b \geq c$ corresponds to a directed edge from node b to node a with a weight of c. If P is achievable, then a set of skews that satisfies P can be found by solving the longest path problem on this directed graph. The clock period P is feasible provided the corresponding constraint graph contains no positive cycles.

This observation was utilized in [DS94] for efficient solution of the skew optimization problem. The optimal clock period is obtained by performing a binary search. At each step in the binary search (for clock period P) the Bellman-Ford algorithm, to be described shortly, is applied to the corresponding constraint graph to check for positive cycles. The procedure continues until the smallest feasible clock period is found. The following skeletal pseudo-code is used to perform the binary search for the optimal clock period.

```
Algorithm MINPERIOD_WITH_OPTIMAL_SKEWS
    Construct the constraint graph;
    P_min = P_low;          ...(Lemma 3.3)
    P_max = P_high;         ...(Lemma 3.4)
    while (P_max - P_min) > ε {
        P = (P_max + P_min)/2;
        if constraint graph has a positive cycle
            P_min = P;
        else
            P_max = P;
    }
```

The binary search can be provably justified since it can be shown that any clock period that is larger than the optimal period is feasible, and that any clock peiod that is smaller is infeasible [DS94, SD96].

The bounds on the the clock period, P_{low} and P_{high}, may be computed as follows. Given a pair of inequalities in Equation (9.9), if we define

$$P_{ij} = \overline{d}(i,j) - \underline{d}(i,j) + T_{setup} + T_{hold} + 2\delta \qquad (9.10)$$

Such a quantity can be defined for any pair of flip-flops, i and j, that are connected by a combinational path. Then a lower bound, P_{low}, is given by

$$P_{low} = \max_{\text{all } (i,j)} P_{i,j} \geq 0. \qquad (9.11)$$

An upper bound, P_{high} is given by

$$P_{high} = \max_{\text{all } (i,j)} (\overline{d}(i,j) + T_{setup} + \delta) \qquad (9.12)$$

9.3.4 Applying the Bellman-Ford algorithm

As stated earlier, each step of the binary search applies the Bellman-Ford algorithm to solve the longest path problem on the constraint graph to determine whether P is feasible or not, and if so, to find the values of the corresponding skews. A brief description of the procedure follows.

The Bellman-Ford algorithm may be applied to solve a single-source longest path problem on a directed graph. Unlike Dijskstra-like algorithms, it can handle edge weights of either sign, but has a higher computational complexity. The pseudocode for the application of the algorithm to a graph $G = (V, E, \mathbf{w})$, where V and E are the set of vertices and edges, respectively, of the graph, and $\mathbf{w}$ is the vector of edge weights, is as follows:

```
Algorithm BELLMAN_FORD
    for each vertex v ∈ V
        xᵥ ← 0;
        predecessor(v) ← ∅;
    for i ← 1 to |V|-1
        for each edge (u,v) ∈ E
            if xᵥ < xᵤ + w(u,v) then
                xᵥ ← xᵤ + w(u,v);
                predecessor(v) ← u;
    for each edge (u,v) ∈ E
        if (xᵥ < xᵤ + w(u,v)) then
            /* A positive cycle exists */
            SATISFIABLE ← FALSE;
        else
            SATISFIABLE ← TRUE;
```

If the skews are initialized to 0, the Bellman-Ford solution achieves the objective of minimizing $|x_{i,max} - x_{i,min}|$ [CLR90]. On a graph with V vertices and E edges, the computational complexity of this algorithm is $O(V \cdot E)$.

In the solution found above, all skews must necessarily be positive, since the weight of each node in the Bellman-Ford algorithm was initialized to zero. Also, in general, the skew at the host node (corresponding to primary inputs and outputs) could be nonzero at the end of these iterations. Since the objective is to ensure a zero skew at the primary input and output nodes, the solution is modified to achieve this. Note that if $[x_1, \cdots x_n]$ is a solution to a system of difference constraints in $\mathbf{x}$, then so is $\mathbf{x}' = [(x_1 + k), (x_2 + k), \cdots , (x_n + k)]$. Therefore, by selecting k to be the negative of the skew at the host node, a solution $\mathbf{x}'$ with a zero skew at the host is found.

9.3.5   Complexity analysis and efficiency-enhancing methods

The number of iterations of the binary search, in the worst case, is $(P_{high} - P_{low})/\epsilon$. The time required to form the constraint graph may be as large as $|f| \cdot |G|$, where $|f|$ is the maximum number of inputs to any combinational stage. In practice, though, it is seen that this upper bound is seldom achieved. Therefore, the iterative procedure above, when carried to convergence, provides the solution to the linear program (9.7) with a worst-case time complexity of

$$O\left(F \cdot E \cdot log_2\left[(P_{high} - P_{low})/\epsilon\right] + |f| \cdot |G|\right) \qquad (9.13)$$

where F is the number of flip-flops in the circuit, E is the number of pairs of flip-flops connected by a combinational path, and P_{high} and P_{low} are as defined in Lemmas 3.3 and 3.4, and ϵ, defined in the pseudocode above, corresponds to the degree of accuracy required.

We caution the reader that the complexity shown above is not a genuine indication of the complexity for real circuits. Firstly, in practical cases, it is seen that $E = O(F)$. Secondly, if the implementation is cleverly carried out, the cost of the computation can be reduced. For instance, it has been shown in [LW83] that such a system of difference constraints, where the graph is "mostly acyclic," a small set of back edges may be identified. During each iteration, the back edges are removed, and the resulting acyclic graph can be solved in topological order (similar to the method in Section 5.2.1) in $O(E)$ time. Next, the weights on the back edges are used to update the node distances, and the iterations continue. It can be shown that at most b iterations are necessary, where b is the number of back edges, resulting in a $O(b \cdot E)$ complexity.

In the outer loop, another approach using back-pointers may be used to reduce the computation. Instead of dividing the interval into half during each step of the binary search, it is possible to use circuit information to reduce the computation and update the search space more efficiently [Szy92]. Suppose that for some value of P, the constraint graph was found to be infeasible. This implies that a positive cycle exists, and each edge on that positive cycle has either a constant weight, if it corresponds to a short-path constraint, or a weight of the form (constant - P), if it is a long-path constraint edge. Given any positive cycle, it is possible to sum up these weights on the edges and update the bound on P so that any value of P that is less than this bound would force this cycle to be infeasible. This reduces the binary search space.

The detection of a positive cycle proceeds as follows. During the Bellman-Ford procedure, a back-pointer `predecessor(v)` is maintained from each vertex v to the predecessor vertex that updated its value; in other words, the edge between these vertices corresponds to an active constraint. After the Bellman-Ford procedure is completed, if the constraint graph is infeasible, a reverse trace is carried out from the vertex v for which it was found that $x_v > x_u + w(u,v)$. This node v must necessarily lie on a positive cycle. The procedure steps back to the predecessor of v, and further back, until v is reached again; at this time, a positive cycle in the graph has been detected. The sum of the weights of all edges on this cycle is calculated as a function of P. As stated in the previous paragraph, this sum is a linear function of P of the form $k_1 - k_2P$; the lower bound on the clock period is updated to k_1/k_2 to ensure that this cycle will not have a positive weight. This procedure can be repeated on every cycle in the graph of predecessor edges; note that the number of edges in such a graph is $O(|V|)$, where $|V|$ is the number of vertices, which means that this can be performed inexpensively. Similarly, if the clock period is feasible, the graph of predecessor edges may be used to update the upper bound on the clock period by exploring the amount by which the clock period could be lowered before the predecessor graph would be altered.

Note that the solution to the problem of clock period minimization corresponds to the maximum P over all positive cycles; it is, in general, not possible to find this maximum in polynomial time since the number of cycles could be exponential in the number of vertices. However, it has been seen that from a practical standpoint, the back-pointer procedure works well in reducing the amount of computation.

9.4   CLOCK SKEW OPTIMIZATION WITH TRANSISTOR SIZING

In the discussion so far, it has been assumed that the delay of each combinational path is fixed. However, the drive strength of gates can be altered by transistor sizing [SK93], thereby changing the combinational path delays. Treating clock skew optimization and transistor sizing as separate optimizations is liable to lead to a suboptimal solution, and a synergistic blend of the two procedures can greatly improve the cost-performance tradeoff for the circuit.

To illustrate this, consider a pair of combinational stages separated by latches, as shown in Figure 9.12(a). We will consider area as the cost function here, but the same idea is valid for other correlated cost functions such as power. If only the sizing transformation were to be applied, the nature of the cost-delay tradeoff curve for each stage would be as shown in Figure 9.12(b). If the timing specification on Stg_1 were stringent, then the optimal sizing solution would lie beyond the knee of the curve, at a point such as point A. Similarly, if the specifications on Stg_2 were loose, then the optimal solution might lie at a point such as C. The cost of the sizing solution would then be [Area(A) + Area(C)].

Now consider a situation where a small skew, S, is applied to the registers between the two stages, allowing Stg_1 to borrow time from Stg_2. This would lead to a looser specification on Stg_1, bringing the solution point from A to B, and a tighter specification for Stg_2, moving its solution from C to D. The cost of the overall sizing solution, [Area(B) + Area(D)], can be seen to be much smaller than the cost with zero skew, even for a small amount of cycle borrowing, since the cost at the solution point A was well above the knee of the curve[2]. Therefore, the judicious use of sizing in conjunction with deliberate skew can deliver solutions of significantly better quality than sizing alone. It is important to perform the two optimizations concurrently, rather than one after another, to fully incorporate the mutual interactions between skew and sizing.

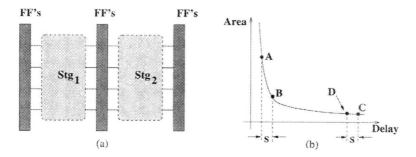

Figure 9.12.   Example illustrating the importance of combining sizing and skew optimization [SSF95].

The remainder of this section presents three techniques for simultaneous optimization of skews and transistor sizes.

9.4.1   A linear programming formulation

The set of long path constraints in the presence of clock skews can be seen from Inequation (9.4) to be linear in the skew variables and the gate delays. If the gate delays could also be represented in terms of linear functions of the transistor sizes, then the set of timing constraints on a circuit would be linear. This idea forms the basis of the linear programming formulation [CSH93, CSH95] discussed in this section.

The delay of a gate is represented as a convex piecewise linear function of its own size and that of its fanout gates. The delay $D(w_k, w_{k,1}, \cdots, w_{k,fo(k)})$ of a gate with size w_k, and with fanout gates of sizes $w_{k,1} \cdots w_{k,fo(k)}$ can be represented using a convex piecewise linear function with q regions as:

$$
D(w, w_{k,1}, \cdots, w_{k,fo(k)}) = \begin{cases} a_1 \cdot w + b_{1,1} \cdot w_{k,1} + \cdots + b_{1,f} w_{k,fo(k)} + c_1 \\ \quad (w, w_{k,1} \cdots w_{k,fo(k)}) \in \text{Region } \mathcal{R}_1 \\ \\ a_2 \cdot w + b_{2,1} \cdot w_{k,1} + \cdots + b_{2,f} w_{k,fo(k)} + c_2 \\ \quad (w, w_{k,1} \cdots w_{k,fo(k)}) \in \text{Region } \mathcal{R}_2 \\ \quad \vdots \\ a_q \cdot w + b_{q,1} \cdot w_{k,1} + \cdots + b_{q,f} w_{k,fo(k)} + c_q \\ \quad (w, w_{k,1} \cdots w_{k,fo(k)}) \in \text{Region } \mathcal{R}_q \end{cases}
$$

$$ (9.14) $$

$$
= \max_{1 \le j \le q} \left(a_j \cdot w + b_{j,1} \cdot w_{k,1} + \cdots b_{j,f} w_{k,fo(k)} + c_j \right)
$$

$$
\forall \, (w_k, w_{k,1} \cdots w_{k,fo(k)}) \in \bigcup_i \mathcal{R}_i. \qquad (9.15)
$$

The second equality follows from the first since $D(w_k, w_{k,1}, \cdots, w_{k,fo(k)})$ is convex.

The variables m_k^i (p_k^i) are introduced to represent the longest (shortest) delay from each flip-flop i to gate k. The introduction of the m and p intermediate variables ensure that the number of constraints depends on the number of gates rather than the number of combinational paths; the number of constraints in the latter case would have been prohibitively large. As before, the skew at flip-flop i is denoted as x_i, while w_k and d_k are the width and delay of gate k respectively. The value of the skew is set to a constant (typically, 0) for primary inputs and outputs. The skew optimization and sizing problem for a general synchronous sequential circuit can now be formulated as follows.

$$\text{minimize} \quad \sum_{k=1}^{\mathcal{N}} \gamma_k \cdot w_k$$

$$
\begin{aligned}
\text{subject to} \quad & d_k \geq D(w_k, w_{k,1}, \ldots w_{k,fo(k)}), & & 1 \leq k \leq \mathcal{N} \\
& w_k \geq \text{Minsize}(k), & & 1 \leq k \leq \mathcal{N} \\
& w_k \leq \text{Maxsize}(k), & & 1 \leq k \leq \mathcal{N} \\
& \text{For all flip-flops } i, & & 1 \leq i \leq \mathcal{L} \\
& \quad x_i + p_k^i \geq x_j + T_{hold} & & 1 \leq j \leq \mathcal{L}, \ k = Fanin(FF_j) \\
& \quad x_i + T_{setup} + m_k^i \leq x_j + P_{spec} & & 1 \leq j \leq \mathcal{L}, \ k = Fanin(FF_j) \\
& \text{For all gates } \ k = 1, \cdots, \mathcal{N} \\
& \quad m_l^i + d_k \leq m_k^i, & & \forall \, l \in Fanin(k) \\
& \quad p_l^i + d_k \geq p_k^i, & & \forall \, l \in Fanin(k)
\end{aligned}
$$

$$(9.16)$$

Here, γ_k is the *area coefficient*, a constant associated with gate k, so that if gate k has size w_k, then its area is $(\gamma_k \cdot w_k)$. In the notation used above, $\mathcal{N}$ is the number of gates, $\mathcal{L}$ is the number of flip-flops in the circuit, and $Fanin(i)$ is the set of gates or flip-flops at the fanin of a gate or flip-flop i.

The problem formulation above is a linear program in the variables d_i, m_i, p_i, s_i and w_i. The entries in the constraint matrix are very sparse, which makes the problem amenable to fast solution by sparse linear programming approaches. The solution technique in [CSH95] also incorporates additional efficiency-enhancing methods that reduce the number of constraints, and employ partitioning techniques to control the problem size with a minimal loss in accuracy.

9.4.2   A convex programming formulation

In [SSF95], a method for combined sizing and skew optimization is presented for acyclic pipelines. As in the case of the linear programming approach just presented, this procedure utilizes the idea of cycle-borrowing using clock skew optimization to relax the stringency of the timing specification on the critical stages of the pipeline. Instead of the convex piecewise linear delay model in the previous approach, this method uses Elmore delays.

The combined problem of sizing and skew optimization is formulated as:

$$\text{minimize} \quad Area$$

$$\text{subject to } x_i + \overline{d}(i,j) + T_{setup} \quad \leq \quad x_j + P + \delta \tag{9.17}$$

$$x_i + \underline{d}(i,j) \quad \geq \quad x_j + T_{hold} + \delta \tag{9.18}$$

$$-X_{max} \quad \leq \quad x_i \leq X_{max} \tag{9.19}$$

where $\overline{d}(i,j)$ and $\underline{d}(i,j)$ are functions of the gate sizes in the circuit, P is the specified clock period, and X_{max} is the maximum allowable skew magnitude. The $Area$ objective function is approximated as the sum of all transistor sizes.

The area of the clocking network is not explicitly included in the formulation for two reasons. Firstly, it is difficult to derive a relation between the skews and the clocking network area. Secondly, the complexity of routing the clock network is such that it is not possible to predict whether a nonzero skew clock tree will necessarily use up more routing resources than a zero-skew tree.

The cost of the clock tree is indirectly modeled through the constraint (9.19). The idea is that since the clock tree is likely to consume routing resources when the skew magnitudes are large, its cost is controlled by limiting the maximum skew magnitude.

Under the Elmore delay model, it can be shown [FD85] that the gate delays are posynomial functions[3] [Eck80] of the gate sizes. A posynomial function in **y** can be transformed into a convex function in **z** using the mapping $y_i = e^{z_i}$. Based on this fact, it was pointed out in [Fis90] that the above optimization problem is a *signomial* programming problem [Eck80] and does not, in general, correspond to a convex program. This makes it difficult to arrive at a good solution to the problem.

It can be seen that each long path constraint is of the form

$$\overline{d}(i,j) + x_i - x_j \leq K, \tag{9.20}$$

where $\overline{d}(i,j)$ is the maximum delay between flip-flops i and j (which is some posynomial function of the transistor sizes), x_i and x_j are clock delays to the source and destination flip-flops, and K is a constant. The left-hand side of this inequality is not a posynomial because of the negative coefficient of x_j. If the logarithmic substitution, $x = e^z$, were performed for each clock skew variable x and transistor size variable w in this inequality, the result would not be a convex constraint.

However, performing the substitution $w = e^z$ for each transistor width w appearing in $\overline{d}(i,j)$, while leaving x_i and x_j alone, results is a convex constraint:

the left-hand side of the inequality is the sum of $\overline{d}(i,j)$, which is convex in the new z variables, and $x_i - x_j$, which is linear (hence convex)[4], in the variables x_i and x_j.

All of the above statements are valid for general sequential circuits, and not just acyclic pipelines. Therefore, it can be concluded that for general sequential circuits, the problem of adjusting both clock skews and transistor sizes to meet long path constraints, while minimizing total gate area, corresponds to a convex optimization program.

Based on these observations, the following approach was proposed to solve this problem in two steps:

1. Neglect the short path constraints (which may be nonconvex [Fis90]) and solve the combined sizing and skew optimization under long path constraints only. This is equivalent to a convex optimization problem and is therefore a unimodal problem, i.e., any local minimum is also a global minimum, and is solved using an adaptation of the TILOS algorithm [FD85].

2. Resolve any short path constraints that are violated at the end of Step 1 by changing the topology of the circuit and adding buffers in an optimal manner using variation of techniques such as [SBS93], described in Section 8.9.

9.4.3 A procedure that incorporates the cost of the clock tree

Both the linear programming and convex programming technique described here control the cost of the clocking network only by placing bounds on the clock skews. An alternative procedure in [XD97] optimizes the cost of the clock tree directly, but uses more approximate models for gate sizing. The optimization problem is set up as

$$\text{minimize } C(T, \mathbf{x}) = \lambda L(T) + \gamma \cdot \Phi(\mathbf{x}) \qquad (9.21)$$

where $C(T, X)$ is a cost function that is dependent on the routing topology T of the clock tree and on the vector of gate sizes $\mathbf{x}$. The functions $L(T)$ and $\Phi(\mathbf{x})$ represent the total wire length of the clock tree, and the power dissipation of the sized gates, respectively, and λ, γ are appropriately chosen weighting parameters. The cost function is optimized using simulated annealing [KGV83].

The algorithm first derives bounds on the relative skew at various sinks. If a pair of latches i and j is connected by combinational logic, then it is possible to derive upper and lower bounds on the value of the difference in skew $x_i - x_j$ using expression (9.9); these are referred to as the positive skew bound, PSB_{ij}, and the negative skew bound, NSB_{ij}. For nonadjacent sinks, no such bound exists. Note that since the expression (9.9) is dependent on the delays, the bounds are derived over all circuit delays that correspond to allowable gate sizes.

The clock tree construction method is based on the DME procedure [KR95], outlined in Section 9.6. Briefly, this is related to the fact that unlike the case of a Euclidean geometry, under a Manhattan geometry, there are several paths between two points that have the same distance. Therefore, when the tapping point is chosen during a bottom-up zero-skew merge, it may lie at a distance $x \times l$ away from one node along *any* of these routes, and the locus of possible tapping points forms a line segment, called a *merging segment*.

In the problem at hand, since the skew is not fixed, but bounded, this permits a variation in the location of the tapping point, and therefore, a variation in the location of the merging segment of the DME procedure. A move in simulated annealing corresponds to changing the position of a merging segment in such a way that the skew bounds are satisfied.

The gate sizing results are stored in the form of a look-up table. The minimum power of each combinational block for various sets of skews is determined and stored in this table. When the clock network is perturbed, the skews are mapped on to the closest set of skew values listed in the table, and the table is accessed to determine the corresponding power dissipation. The cost of this configuration is evaluated and fed into a simulated annealing loop that finds the optimal solution to the problem.

9.5 TIMING ANALYSIS OF SEQUENTIAL CIRCUITS FOR SKEW SCHEDULING

The optimization in Section 9.4.2, and indeed, many other procedures that work with clock skew, requires the detection of violations of the clocking constraints. This may be carried out by using a modification of CPM for delay estimation that is generalized to handle sequential circuits [SSF95].

The procedure for timing analysis requires the determination of the optimal skews and the arrival times at all gate outputs. Since the constraint graph for an acyclic pipeline is acyclic, and since timing analysis involves the solution of the longest path problem on this constraint graph, the complexity can be reduced substantially by processing the vertices in topological order [CLR90]. Specifically, the timing analysis procedure is equivalent to applying CPM, described in Section 5.2.1, to the graph, with gates being represented in the normal way, and flip-flops being represented by blocks with "delay" values of $T_{setup} - P - \delta$, where T_{setup} is the setup time for the flip-flop and P is the applied clock period, and δ is the uncertainty in the clock skew[5].

We point out here that the reference point for the arrival time in each combinational block is the zero skew. For example, an arrival time of a at the output of a flip-flop implies that the clock signal at the flip-flop has been skewed by a. If this flip-flop fans out to a single inverter with a delay of d, then the arrival time at the output of this inverter would be $a + d$. Since skews may be either positive or negative, it can be seen that arrival times may have either sign.

Consider Figure 9.13(a). We symbolically show the path from FF_i to FF_j as being represented by a single gate with delay $G_{maxdelay}$, which corresponds to the maximum combinational delay between these flip-flops. From (9.17) we

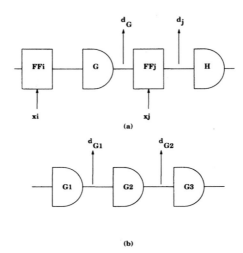

(a)

(b)

Figure 9.13. Modifying CPM to analyze sequential circuits for clock skew scheduling [SSF95].

have

$$x_i + G_{maxdelay} + T_{setup} \leq x_j + P + \delta \qquad (9.22)$$

where x_i is the latest arrival time at the input of G, x_j is the latest arrival time at the input of gate H. Writing $x_i + G_{maxdelay}$ as d_G, the latest arrival time at the input of FF_j, we have the following constraint

$$x_j - d_G \geq T_{setup} - P - \delta \qquad (9.23)$$

For a regular combinational gate, as shown in Figure 9.13(b), if d_{G1} and d_{G2} are the latest arrival times at the input of $G2$ and $G3$, then we have the difference constraint

$$d_{G2} - d_{G1} \geq delay(G2) \qquad (9.24)$$

From the two difference constraints given above it can be seen that FF_j behaves like a gate with a delay of $T_{setup} - P - \delta$.

It is noteworthy that CPM is valid if the circuit is acyclic; if not, techniques such as those used in [LW83] will have to be applied for analysis. Alternatively, for optimization purposes, a cyclic graph can be forced to be acyclic using the techniques in [BSM94], with additional constraints imposed at points where the cycles are broken.

The process of calculating the optimal skews falls out as a natural consequence of this procedure: if the timing requirements are met, then the arrival time at the output of a flip-flop, as calculated by CPM, is a valid value of the skew to be applied to that flip-flop.

9.6   WAVE PIPELINING ISSUES

The treatment in [GLC94] provides an excellent illustration of the idea of wave pipelining, and is summarized here. A conventional combinational circuit and its associated timing are shown in Figure 9.14(a). The data flows out of the input registers on the onset of the clock, and propagates through the logic, reaching the output registers in time for the onset of the next clock.

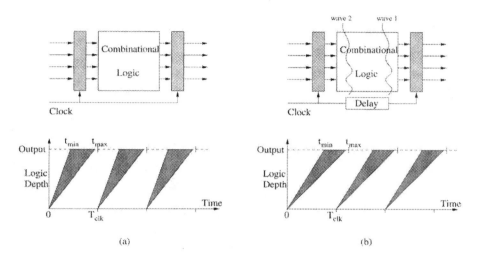

Figure 9.14.   Illustrating the idea of wave pipelining [GLC94].

In contrast, in a wave pipelined system, the propagation time for the data through the logic may be larger than the clock period. As an example, in Figure 9.14(b), the clocks at the input and output register are skewed. The data that leaves the input registers reaches the output registers after more than one clock period. During this time, the propagation of the second set of data from the input registers has commenced at time T_{clk}. Therefore, there is a period of time when two different signals corresponding to two different clock periods are flowing through the combinational logic, corresponding to two "waves." To ensure correct operation of the circuit, it is imperative to ensure that the two waves do not collide. The advantage of wave pipelining over conventional pipelining is that it may not require the use of as many registers, leading to potential savings in hardware and in the overhead of setup and hold times at each register. Moreover, the overhead of clock distribution is also reduced when the number of registers is reduced.

The critical issue in wave pipelining is to ensure that no two waves collide during transmission. The work in [JC93] proposed an alteration of the clock skew formulation of Fishburn [Fis90] to incorporate constraints that enforce this requirement; these constraints are referred to as logic signal separation constraints. Figure 9.15 shows two flip-flops, k and l, and a set of combinational logic gates, q, g and s. Node q is any predecessor gate of g such that flip-flop

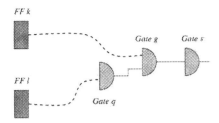

Figure 9.15. An example illustrating logic signal separation constraints [GLC94].

l lies in its fanin cone, and gate s is any successor of node g. Let the skews at the flip-flops k and l be x_k and x_l, respectively. The logic signal separation constraint under a clock period P is stated mathematically as follows:

$$x_k + \text{delay}(k, s) \leq P + x_l + \text{delay}(l, q) \tag{9.25}$$

The left hand side of this equation represents the arrival time of the signal from the current wave at the output of the successor of g. This arrival implies that the value at g is no longer important for the current wave. The right hand side is the arrival time at the output of q during the next wave. Simply stated, the inequality says that it is essential for the inputs of g to be stable for as long as its output influences the current wave. In case g has multiple predecessors and successors, the relationship (9.25) must hold between each predecessor-successor pair.

Earlier in this chapter, several techniques for combined sizing and skew were described. None of them explicitly considers the logic signal separation constraints in the form in which they were published. However, it is relatively easy to incorporate them into the formulations. For the linear programming formulation, the output arrival time at each gate is available as a variable, and it is an easy matter to add constraints of the form (9.25) at $m_s \leq p_q + P$. For the convex programming formulation, this constraint can be checked during the application of CPM, and any violation can be flagged as a constraint violation to be reconciled by altering the sizing and/or skew solution.

9.7 DELIBERATE SKEWS FOR PEAK CURRENT REDUCTION

In synchronous circuits, all of the flip-flops switch at the latching edge of the clock. This simultaneous switching of flip-flops requires a large switching current, which in turn creates a significant voltage drop on the power distribution network, creating the need for multiple pins to distribute the power supply. In [VHBM96] deliberate clock skews are used to reduce the dynamic transient current drawn from the supply pins. This reduction in the transient current reduces the number of supply pins required by the chip, and therefore, its packaging cost.

The objective of introducing skews at the flip-flops is to ensure that all of them do not switch at the same time and hence, do not draw current at the same time; this serves to reduce the peak current. The problem is modeled as an integer linear program that minimizes the peak current. Each flip-flop may have one of n possible skew values. Correspondingly, for each flip-flop a set of n binary variables is used, of which exactly one variable is permitted to be 1, and the skew of the flip-flop corresponds to this variable. The clock period of the circuit imposes constraints on the possible skew value for each flip-flop, these timing constraints are incorporated in the linear program. Experimental results presented in [VHBM96] show a significant reduction in the ground bounce using this technique. However, the use of an integer linear programming formulation restricts the technique to system level design with few tens of modules. It is possible that the application of heuristics to this technique may permit it to be extended to larger designs, at the cost of sacrificing optimality.

A method for heuristic minimization of peak current by using clock skew is presented in [VBBD96, BVBD97]. Experimental results on circuits with up to 550 flip-flops indicate an average reduction of about 30% in the peak current. This method only minimizes the current peak directly caused by clock edges using a genetic algorithm. The current waveform is approximated as a triangle to allow efficient calculation of the total peak current in the circuit. The genetic algorithm based solution results in a required skew at each flip-flop. Since clock skew control for individual flip-flops is difficult, the flip-flops are clustered into a user specified number of clusters, and each flip-flop in a cluster has the same skew. A heuristic is used to attempt to cluster the flip-flops in such a way that there is a minimal loss in optimality due to this simplification.

9.8   SUMMARY

The utility of deliberate skews for optimizing the performance of VLSI circuits has been demonstrated, and algorithms for performing skew scheduling and period minimization have been presented. Deliberate skews can also be used in conjunction with other timing optimization strategies and for minimizing the peak current in the supply network.

Until recently, there has been a great reluctance to alter the clock network and attempt a nonzero-skew solution. However, recently, an increasing number of designers have been willing to utilize skews for performance enhancement. Small amounts of skews can easily be provided by making minor changes in the sizes of the final buffers in the clock tree that feed the clock sinks, and altering skews in this manner is a relatively painless manner in which the optimization could be applied. For larger skew magnitudes, a more careful design of the clock network is essential; for high-performance applications, the gains outweigh the costs of this effort.

Notes

1. In subsequent generations of the processor, the regional clocks are provided by tree-based structures [TDL03].

2. In this example, the change in the cost of the clock network has been neglected. It should be noted that it is easy to provide small skews by making minor changes in the sizes of the buffers driving the leaf-level nodes, and it possible to do this cheaply for small skew magnitudes.

3. A posynomial is a function g of a positive variable $\mathbf{w} \in \mathbf{R}^n$ that has the form $g(\mathbf{w}) = \sum_j \gamma_j \prod_{i=1}^n w_i^{\alpha_{ij}}$ where the exponents $\alpha_{ij} \in \mathbf{R}$ and the coefficients $\gamma_j > 0$. Roughly speaking, a posynomial is a function that is similar to a polynomial, except that (a) the coefficients γ_j must be positive. (b) an exponent α_{ij} could be any real number, and not necessarily a positive integer, unlike the case of polynomials.

4. This concept can be generalized to any optimization problem that can be divided into two separate classes $w_1, \cdots, w_n$, and $x_1, \cdots, x_m$, with each constraint is of the form

$$P(w_1, ..., w_n) + C(x_1, ..., x_m) \le K, \tag{9.26}$$

where P is a posynomial function, C is a convex function, and K is a constant. Such a problem can be transformed into a convex program by performing the substitution $w_i = e^{z_i}$ while leaving the x_i variables alone.

5. Note that the word "delay," when used in reference to flip-flops, is applied in a loose sense here. The "delay" of a flip-flop as defined here is not the propagation delay of the gates within the flip-flop, but is a mathematical tool that can be used to apply CPM to a sequential circuit to check for delay violations in the presence of clock skews.

10 RETIMING

10.1 INTRODUCTION TO RETIMING

Retiming is a powerful sequential circuit optimization technique for improving the performance of sequential circuits. The concept of retiming is the notion of moving storage devices across memoryless computational elements to improve the performance without changing the input-output latency. Although retiming can operate on gate level netlists, or on higher-level abstractions such as data flow graphs, communication graphs, and processor schedules, our treatment will focus on circuit-level optimizations.

At the circuit level, the storage devices, or registers, may either be edge-triggered flip-flops or level-sensitive latches, commonly referred to as FF's and latches, respectively. The computational nodes at the circuit level are typically combinational gates. Retiming moves registers across gates without changing the number of registers in any cycle or on any path from the primary inputs to the primary outputs, thereby preserving the input-output latency of the circuit. Since retiming does not directly affect the combinational part of the circuit, the circuit behavior remains unchanged. However, since retiming can change the boundaries of combinational logic, it has the potential to affect the results of combinational synthesis techniques that are applied to the sequential circuit.

This chapter begins with an overview of the retiming procedure. Next, the assumptions and models common to many retiming algorithms are presented, followed by a description of the Leiserson-Saxe theory for retiming edge-triggered circuits is presented. Finally, we discuss fast and efficient methods for retiming large edge-triggered and level-clocked circuits to minimize their clock period.

10.1.1   Types of retiming

Retiming may be performed to improve the circuit behavior with respect to several possible objective functions, some of which are outlined below.

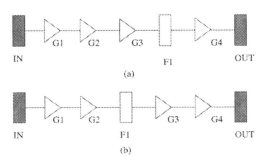

Figure 10.1.   The effect of retiming on the clock period: assuming unit delays and zero setup times, the circuit in (a) has a clock period of 3 units, which is reduced to 2 units in (b). The two circuits are retimed versions of each other.

Clock period   Algorithmically, the simplest objective function used in retiming is the minimization of the clock period. Since the clock period in a circuit with edge-triggered FF's is given by the maximum combinational delay, the FF's may be relocated to reduce the clock period. For the circuit shown in Figure 10.1(a), with unit delay gates and edge-triggered FF's, the clock period is 3 time units. If we relocate register F1 from the output of gate G3 to its input, we obtain the circuit in Figure 10.1(b), with a clock period of 2 units. Notice that the input-output behavior is left unchanged by retiming since the output is produced after two clock cycles in both the original and the retimed circuit. Thus, relocating registers can reduce the clock period of a circuit, and retiming can be used to relocate registers with the objective of minimizing the clock period. A retiming that minimizes the clock period of a circuit is termed a *minimum period retiming*. Retiming a circuit to achieve a specified target clock period is a special case of minimum period retiming, and is often called specified-period retiming.

Area   Since retiming does not affect the combinational part of the circuit, the area overhead of the combinational logic remains constant under retiming. The method may, however, affect the overall area of the circuit since it may alter the number of registers in the circuit.

Two retimed versions of the same circuit could have the same input-output behavior and clock period, but could use a different number of registers, as

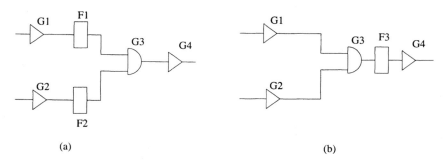

Figure 10.2. The effect of retiming on the number of registers: the two circuits are retimed versions of each other, but (a) has more registers than (b).

illustrated in Figure 10.2. The circuit in Figure 10.2(a) requires two registers, while that in Figure 10.2(b) requires only one register.

Therefore, one could apply retiming with the object of minimizing the number of registers in the circuit, while leaving the input-output latency unchanged. This may be done without any constraint on the clock period of the resulting circuit, or subject to a target clock period. The former is called unconstrained minimum area retiming while the latter is referred to as constrained minimum area retiming or simply *minimum area retiming*. In practice, minimum area retiming is a more useful form of the retiming transformation than minimum period retiming.

Power The power dissipated in a circuit depends on the product of the switching activity and the load capacitance at the output of a gate, summed over all gates. Since registers can filter out glitches, altering their locations can affect the switching activity at gate outputs; moreover, it can also alter the load capacitance seen by the gates. Thus, retiming can change the power dissipation of a circuit, and an appropriately chosen retiming may be used to optimize the power by placing registers at nodes with high switching activity values and high capacitive loads.

Testability The relocation of registers can change the state encoding in sequential circuits, thus affecting the test generation time and the number of redundant faults. The repositioning of registers also affects the length of the scan chains required for partial or full scan designs. Retiming can, therefore, be used to improve the testability of sequential circuits.

Quality of logic optimization Most logic optimization techniques operate on combinational logic blocks separated by register boundaries. Hence, changing these register boundaries by retiming the registers affects the quality of results obtained by logic optimization.

A brief survey of publications describing these research activities is presented in Section 10.2, and in the next few chapters, we will describe these methods in varying degrees of detail.

10.1.2 Types of circuits

Algorithms for retiming a circuit must address the specific requirements of a circuit, and the clocking discipline used. Four major classes of circuits are described below.

Edge-triggered circuits In an edge-triggered circuit, the clock period is given simply by the largest combinational delay. The first publications on retiming concentrated on handling this class of circuits.

Level-clocked circuits Level-sensitive latches are, by definition, transparent during the period when the clock signal is active; this transparent nature gives level-clocked circuits the potential to operate at a faster clock period. These circuits require less area than their edge-triggered counterparts, not only because the cycle stealing reduces the amount of transistor sizing required to meet the timing goals, but also because individual latches require a smaller amount of area than edge-triggered FF's. Unfortunately, the analysis of level-clocked circuits is more complicated than edge-triggered circuits, and hence algorithms for finding an optimal retiming can be computationally expensive.

Control logic Control logic involves an implementation of Finite State Machines (FSM's), and hence the registers in control circuitry are associated with the FSM states. Retiming alters the locations of these registers, and consequently, the state encoding of the FSM. Thus, issues regarding safe replaceability become important. In circuits that have a meaningful initial state, it is important to find a retimed circuit with an equivalent initial state; many retimings of a circuit that are otherwise valid may not have equivalent initial states. To see this, consider the circuit in Figure 10.3(a). If we wish to move FF A and B across gate G1 (to FF C in Figure 10.3(b)), we must find an initial value of FF C that is equivalent to the initial values of FF A and B. If FF A and B have conflicting values, no such equivalent initial value exists at FF C. Thus, additional constraints must be imposed to ensure the presence of an equivalent initial state while retiming control logic.

FPGA's Field Programmable Gate Arrays (FPGA's) present requirements that are different from those in conventional combinational logic. For example, in LUT-based FPGA's the amount of logic is dependent on the number of inputs, and not on the complexity of the logic. Further, since FPGA's have limited resources with memory elements at fixed locations, extra constraints are placed on the movement of memory elements during retiming. The issue of combined synthesis with retiming shows the maximal gains in FPGA optimization.

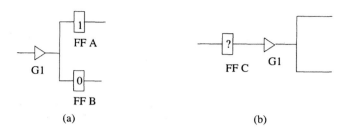

Figure 10.3. The problem of initial state equivalence in reverse retiming: the states in (a) do not have a corresponding state in (b).

10.2   A BROAD OVERVIEW OF RESEARCH ON RETIMING

Since retiming was introduced by Leiserson and Saxe [LRS83, LS91], a significant amount of research has been carried out on retiming both in academia and in industry (for example, at IBM, Philips and Synopsys). In this section, we present a brief literature survey of retiming-related research. For a good introduction and overview of retiming, the reader is also referred to [She97] and to Section 9.3.1 of [De 94]. In the remainder of this chapter, more detailed descriptions of these issues will be presented.

Edge-triggered circuits Leiserson and Saxe introduced algorithms for minimum period and minimum area retiming of edge-triggered circuits [LRS83]. The circuit is represented by a graph and polynomial time algorithms are presented. The major contribution of this work is in formulating the theory of retiming, rather than in presenting experimental implementations. The minimum period retiming problem is solved by performing a binary search for the best clock period. The feasibility of a given clock period is checked by a Bellman-Ford-like relaxation algorithm, and the minimum area problem is formulated as a linear program (LP). This LP is shown to be the dual of a mincost network flow problem. Details of this approach, which we call the "LS approach," are provided in [LS91] and described briefly in Section 10.5.1.

Shenoy and Rudell presented an efficient and clever implementation of the LS algorithms in [SR94]. Their main contributions include reducing the memory requirements from $O(|G|^2)$ to $O(|G|)$, where $|G|$ is the number of gates in the circuit, and the use of back-pointers to speed up the feasibility check during the binary search for minimum clock period. At about the same time, a technique for reducing the number of constraints in the minimum area LP was presented in [vVA+94].

The ASTRA algorithm [Deo94, DS95, SD96] exploited the retiming-skew equivalence for fast minimum period retiming. ASTRA first finds a minimum period achievable by skew optimization, and then translates these skews into retiming. Circuits with 20,000 gates were shown to be retimed in under two minutes.

The Minaret algorithm [MS97a, MS98b] modified ASTRA to efficiently obtain bounds on the retiming variables, and used them to significantly reduce the size of the minimum area LP. This enabled Minaret to perform minimum area retiming on circuits with 50,000 gates in under 15 minutes. The work in [SSP99] extends the formulation above to handle short path constraints as well, permitting similar algorithms to be used for this purpose.

Level-clocked circuits A signal that flows through a level-sensitive latch during its transparent phase can initiate the computation of the next combinational stage before the beginning of the next clock cycle; this phenomenon is called *cycle stealing*. Due to cycle stealing, level-clocked circuits have the potential to operate faster, and require less area. Algorithms to retime single phase level-clocked circuits are presented in [SBS91], and techniques based on the LS model for retiming multi-phase level-clocked circuits were presented in [ILP92, LE92, LE94]. TIM [PR93b] is a comprehensive timing analysis and optimization CAD tool for level-clocked circuits that is available in public domain, and has been used to empirically compare the performance of edge-triggered and level-sensitive circuits in [PR93a]. The work in [IP96] presented a polynomial-time algorithm for pipelining two-phase, level-clocked circuits under a bounded delay model.

The ASTRA and Minaret algorithms for edge-triggered circuits have been extended to level-clocked circuits in [MS96] and [MS98a] respectively. This enables minimum period and minimum area retiming of level-clocked circuits with tens of thousands of gates in very reasonable time.

Retiming with equivalent initial states Traditional retiming algorithms do not pay any regard to initial states or power-on states of circuits, and are not very useful for control logic. Control logic usually has meaningful initial states, and any useful retiming must also find a new initial state for the transformed circuit that is equivalent to the initial state of the original circuit.

A method for minimum period retiming with equivalent initial states was presented in [TB93], using only the so-called forward retiming moves. In some cases, this approach may require modifications in the combinational part of the circuit. An efficient technique for performing these modifications is presented in [SMB96], and a method for preserving synchronizing sequences after retiming is proposed in [MSM04]. An alternative approach, termed reversed retiming [ESS96, SSE95], uses a minimum number of reverse (backward) retiming moves, and precludes the need for any modifications to the combinational part of the circuit. An approach to minimum area retiming that maintains initial states is presented in [MS97b].

Low power Retiming can affect the power consumption of a circuit since it alters the amount of switching in a circuit, as also the fanout capacitance driven by various elements in the circuit. A mechanism for reducing power by retiming was presented in [MDG93]; it places FF's on interconnects with high switching activity. The approach in [LP96] presented algorithms to reduce power by retiming only one phase in a two-phase circuit. The advantage of retiming only one phase is that it preserves the testability of the circuit. A similar approach is taken in [SSWN97] to reduce power in DSP designs.

Testing and verification Retiming can be used both to improve the testability of a circuit, and as an aid to automatic test generation. In the former case, circuit is actually implemented in the retimed form, while in the latter case the retimed circuit is merely used by the test generator to generate test vectors, but the original circuit is implemented. Various researchers have worked on characterizing the effects of FF relocation on the redundancy of faults [DB96, DC94, YKK95a, YKK95b], and the corresponding effects on ATPG run time [MEMR96]. In [EMRM95, EMRM97], it was shown that retiming preserves testability with respect to a single stuck-at-fault test set by adding a prefix sequence of a pre-determined number of arbitrary input vectors.

Retiming has been used to improve testability [DC94] by attempting to convert sequential redundancies to combinational redundancies. Retiming may also be employed for reducing test lengths in scan-based designs [HKK95, HKK96, KT94], and for improving pseudo-exhaustive built-in self test (BIST) [KTB93, LKA95a, LKA95c, LKA95b].

The work in [SPRB95] shows that while an accurate logic simulation may distinguish a retimed circuit from the original circuit, a conservative three-valued simulator cannot do so. Techniques for verification of retimed circuits are presented in [HCC96, HCC97, RSSB98].

Enhancements to retiming The traditional retiming approaches assume the gate delays to be fixed, and all FF delays to be equal. Since these are only approximations, much effort has been spent in incorporating improved delay models into retiming. Delay models that incorporate clock skews, register delays, etc., are presented in [SF94, SFM93, SFM97, LP95a, LP95b], while [KO95] presents retiming under a variable delay model. Issues related to integrating retiming with placement are addressed in [TST⁺98, NK01].

Retiming has also been extended to handle multiple clocks and registers with enable ports in [LVW97, Mar96], and to handle gated clocks and precharged structures [Ish93]. Various efforts have been made to combine retiming with other logic synthesis techniques, e.g., [BC96, LKW93, Lin93, MSBS90, MSBS91, Pan97, PR91, AB98, BK01]. Architectural retiming [Has97] modifies the combinational part of a circuit to increase the number of registers on a critical cycle or path without increasing the perceived latency.

Other applications Retiming has been used in numerous other applications: during the technology mapping step in FPGA synthesis [CW94, CW96, PL95, TSS92, WR93], to improve circuit partitioning [LSC⁺93], for scheduling in high level synthesis [CS93a, WBL⁺94], in multiprocessor scheduling [CS92], for system level throughput optimization include [MH91, DPP95], etc. Retiming has also been used extensively in DSP applications [Dun92, Geb93, PS96, PSB96, PR92].

10.3   MODELING AND ASSUMPTIONS FOR RETIMING

Any detailed presentation of material on retiming necessarily requires a section explaining the models and preliminaries that are utilized by most algorithms. In this section, we will present a graph-theoretic foundation that will be utilized later in our exposition on retiming.

In the ensuing discussion, we will utilize the digital correlator circuit from [LS91], shown in Figure 10.4(a), to motivate the procedure of applying retiming by demonstrating the performance improvements on this example. The design consists of two types of functional elements: adders (denoted by a '+' symbol) and comparators (denoted by a '=' symbol). The boxes between the comparators are registers that act to shift the data. The delay of an adder is 3 units and that of a comparator is 7 units. Although this design is functionally correct, it has poor timing characteristics.

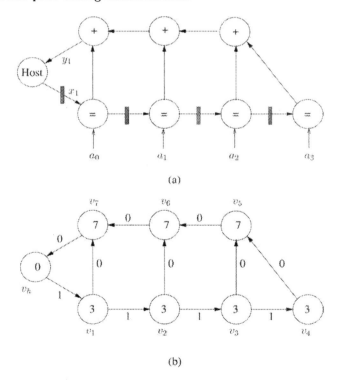

(a)

(b)

Figure 10.4.   (a) A correlator circuit, and (b) a graph that represents the circuit [LS91].

For the purposes of applying a retiming algorithm, it is useful to abstract the problem from the circuit level into a directed graph representation, $G(V, E, d, w)$, where V is the set of vertices and E the set of edges. Each vertex $v \in V$ models a functional element with a fixed propagation delay, $d(v)$, that remains unaltered during the retiming process. A special vertex called the *host* is introduced in the graph, with outgoing edges connected to all primary inputs of the circuit, and incoming edges from all primary outputs to the host vertex. The host vertex is assigned a propagation delay of zero, i.e., $d(host) = 0$. The directed edges model the interconnections between functional elements. Each edge $e_{uv} \in E$ connects an output of functional element u to an input of a functional element v, with a weight of $w(e_{uv})$ that corresponds to the number of registers between u and v. The graph corresponding to the correlator circuit in Figure 10.4(a) is shown in Figure 10.4(b).

As an informal definition, a retiming move in a circuit is caused by moving all of the memory elements at the input to a combinational block to all of its outputs. A sequence of such movements is referred to as a retiming of the circuit.

Equivalently, working on the graph representation of the circuit, a retiming is an integer-valued vertex-labeling of the vertices, $r : V \rightarrow \mathbf{Z}$, where $\mathbf{Z}$ is the set of integers. The weight of an edge e_{uv} after retiming, denoted by $w_r(e_{uv})$, is given by

$$w_r(e_{uv}) = r(v) + w(e_{uv}) - r(u) \qquad (10.1)$$

The retiming label, $r(v)$, for a vertex v represents the number of registers that have been moved from its outputs to its inputs. Retiming can, therefore, also be viewed as an assignment of a lag value, $r(v)$, to every vertex v in the circuit. A path originating at vertex u and terminating at vertex v is represented as $u \leadsto v$. For a path $p : u \leadsto v$, one may define its weight, $w(p)$, as the sum of the weights on the edges on p, and its delay, $d(p)$, as the sum of the weights of the vertices on p. A path with $w(p) = 0$ corresponds to a purely combinational path with no registers on it. Therefore, the clock period, P, can be calculated as

$$P = \max_{\forall\, p | w(p)=0} \{d(p)\} \qquad (10.2)$$

The Leiserson-Saxe method presents a systematic technique for computation of the retiming labels, where the problem is formulated as a *Mixed Integer Linear Program* (MILP). An important concept used in the approach is the notion of the W and D matrices. The matrices are defined for all pairs of vertices (u, v) such that there exists a path, $p : u \leadsto v$, that does not include the host vertex. The formal definition of the matrices is as follows:

$$W(u, v) = \min_{\forall\, p : u \leadsto v} \{w(p)\} \qquad (10.3)$$

$$D(u, v) = \max_{\forall\, p : u \leadsto v \text{ and } w(p)=W(u,v)} \{d(p)\} \qquad (10.4)$$

In plain English, $W(u, v)$ denotes the minimum latency, in clock cycles, for the data flowing from u to v, and $D(u, v)$ gives the maximum delay from u

to v over all paths with that minimum latency. The (W,D) pairs are used to generate constraints, and the essential reason why $D(u,v)$ is important is that if all memory elements between u and v are removed during retiming, then the delay between the two vertices on the path corresponding to $D(u,v)$ will be the maximum purely combinational delay between the two vertices. In the next section, we will use this information to ensure that if $D(u,v) > P$, the clock period, then the value of $W(u,v)$ after retiming is at least 1. The important point here is that the delays on all but the minimum weight path are irrelevant.

The W and D matrices can easily be computed as follows. Each edge, e_{uv}, is reweighted with the ordered pair $(w(e_{uv}), -d(u))$, and the all-pairs shortest paths are computed. Any standard procedure for finding all-pairs shortest paths, e.g., the Floyd-Warshall method, or Johnson's algorithm [CLR90], may be used, with the following change that allows an ordered pair of weights to be correctly interpreted. During the relaxation process, comparisons between ordered pairs are performed in lexicographic order, i.e., amongst two pairs, $p_1 = (a_1, b_1)$ and $p_2 = (a_2, b_2)$, we say that $p_1 < p_2$ if $(a_1 < a_2)$ or $(a_1 = a_2$ and $b_1 < b_2)$. At the end of this computation, if the shortest-path weight between two vertices u and v is (x, y), then $W(u,v) = x$ and $D(u,v) = d(v) - y$.

10.4 MINIMUM PERIOD OPTIMIZATION OF EDGE-TRIGGERED CIRCUITS

The algorithms presented in this section are all of polynomial time complexity. We will first present algorithms that are rooted in the techniques presented by Leiserson and Saxe [LS91] for retiming edge-triggered circuits, and follow these by discussing how they may be applied efficiently in practice. Finally, we will describe a method that applies the retiming-skew relationship to find an efficient minimum period retiming solution.

In each of these methods, the minimum period obtainable under retiming is calculated by performing a binary search over all possible clock periods. At each step in the binary search, an attempt is made to retime the circuit for the current value of the clock period. The smallest period for which retiming succeeds is returned as the best clock period.

10.4.1 Leiserson-Saxe-based algorithms for minimum period retiming

Any retiming solution must satisfy the following two inequalities

$$r(u) - r(v) \leq w(e_{uv}) \qquad \forall e_{uv} \in E$$
$$r(u) - r(v) \leq W(u,v) - 1 \quad \forall D(u,v) > P \qquad (10.5)$$

The first constraint ensures that the weight of each edge, e_{uv} (i.e., the number of registers between the output of gate u and the input of gate v), after retiming is nonnegative, i.e., $w_r(e_{uv}) \geq 0$. We will refer to these constraints as *circuit constraints*. The second constraint ensures that after retiming, each path whose delay is larger than the clock period has at least one register on it. These constraints, being dependent on the clock period, are often referred to as *period constraints.*

For a fixed clock period, the set of constraints (10.5) correspond to a system of difference constraints. This system can be represented by a constraint graph that may be solved by applying the Bellman-Ford method [CLR90]. If the system of constraints is infeasible, this will be indicated by the Bellman-Ford procedure. The smallest clock period for which the constraints are feasible is the solution to the minimum period problem.

This notion leads to the first algorithm, OPT1 [LS91]. As mentioned earlier, in the outer loop, a binary search in performed on the value of the clock period, and for each clock period, the Bellman-Ford algorithm is applied to check for feasibility of the constraint set. The binary search for the minimum value of P is justified by the fact that the feasible values of P form a continuous interval. This is due to the fact that the system (10.5) of inequalities is linear in all variables, making the solution space convex. Consequently, the projection of the feasible region on P is convex in one dimension, i.e., it is an interval. The OPT1 algorithm requires the calculation of the W and D matrices, and is expensive in terms of computation time ($O(V^3 \log V)$) and memory space ($O(V^2)$).

An alternative $O(|V||E|\log|V|)$-time algorithm from [LS91] is a more practical option for obtaining a retiming. The main routine FEAS is invoked during the binary search to check whether a specified clock period, P, is feasible or not. It proceeds by calling a subroutine, CP, that calculates the clock period of the circuit by systematically identifying the purely combinational path that has the largest delay. For any gate whose delay from an FF exceeds P, the retiming label, r, is incremented.

Algorithm FEAS
Given a synchronous circuit $G = \langle V, E, d, w \rangle$, and a desired clock period c, return a retiming r of G such that the clock period of the retimed circuit $\Phi(G_r) \geq P$.
{
  1. For each vertex $v \in V$, set $r(v) \leftarrow 0$.
  2. Repeat the following $|V| - 1$ times
      2.1 Compute the graph G_r with the existing values of r.
      2.2 Run Algorithm CP on the graph G_r to determine $\Delta(v)$
          for each vertex $v \in V$.
      2.3 For each v such that $\Delta(v) > P$, set $r(v) \leftarrow r(v) + 1$.
  3. Run Algorithm CP on the circuit G_r. If $\Phi(G_r) > P$,
      then no feasible retiming exists. Otherwise, r is the
      desired retiming.

}

Algorithm CP
This algorithm computes the clock period $\Phi(G)$ for a synchronous
circuit $G = \langle V, E, d, w \rangle$.
{
1. Let G_0 be the subgraph of G that contains precisely those
    edges e such that the register count $w(e) = 0$.
2. Perform a topological sort on G_0, totally ordering
    its vertices so that if there is an edge from vertex u
    to vertex v in G_0, then u precedes v in the total order.
3. Go through the vertices in the order defined by the
    topological sort.
    On visiting each vertex v, compute $\Delta(v)$ as follows:
    a. If there is no incoming edge to v, $\Delta(v) \leftarrow d(v)$.
    b. Otherwise, $\Delta(v) \leftarrow d(v) + \max\{\Delta(u) : u \xrightarrow{e} v$ and $e \in G_0$.
4. The clock period $\Phi(G)$ is $\max_{v \in V} \Delta(v)$.
}

The advantage of this method is that it does not require the explicit calculation of the W and D matrices, thereby reducing the memory overhead. Step 2
of FEAS can be shown to be equivalent to applying the Bellman-Ford algorithm
on the graph G.

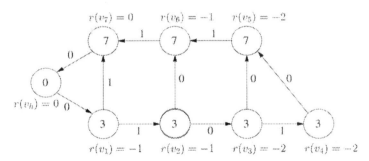

Figure 10.5.   The retimed graph for the correlator, with the r values shown next to each
vertex, and with the updated edge weights [LS91].

The application of either of these retiming procedures may be used to assign the r labels to each vertex in the graph corresponding to the correlator
circuit shown in Figure 10.4(a). For the unretimed circuit, the clock period
is 24 units, which corresponds to the sum of the propagation delays along the
longest register-free path, $v_4 \rightarrow v_5 \rightarrow v_6 \rightarrow v_7$. The graph obtained on applying minimum period retiming to Figure 10.4(b) is shown in Figure 10.5.
The r labels describe the manner in which flip-flops are added or removed
from the original graph. For instance, the edge (v_1, v_7) has one more register since $r(v_7) - r(v_1) = 1$ (see relation (10.1)). Since retiming preserves the
input-output latency of the circuit, the original and the retimed circuits corre-

sponding to graphs in Figure 10.4(b) and Figure 10.5, respectively, must have the same input/output behavior. However, the computations performed by v_1 in Figure 10.4(b) lead by one clock tick the same computations performed by v_1 in Figure 10.5. This is the physical meaning of the fact that the vertex v_1 is assigned a lead of one clock tick or equivalently, a lag $r(v_1) = -1$. The clock period of the retimed circuit is 13 time units, corresponding to the sum of the propagation delays along the longest path, $v_2 \rightarrow v_3 \rightarrow v_5$.

Implementational notes.   An efficient implementation of the minimum period retiming algorithm was presented in [SR94], using a predecessor heuristic. The procedure is similar to the method of [Szy92], presented in Section 9.3.5, and its adaptation to this problem is presented here. The predecessor heuristic maintains a predecessor vertex pointer, denoted by pred(), for each vertex, which is initialized to the empty set $\emptyset$. When $\Delta(v)$ is computed for each vertex v, a reference to a vertex u is stored, where u satisfies the property that there exists a zero-weight path $p : u \rightsquigarrow v$ and $\Delta(v) = d(p)$. If $\Delta(v) > P$, then the procedure sets pred(v) = u.

Consider a cycle in the predecessor subgraph with vertices $u_1, \cdots, u_k = u_0$, i.e., pred(u_i) = u_{i-1}, $i = 1, \cdots, k^1$. Let $p_{i-1,i}$ denote the path $u_{i-1} \rightsquigarrow u_i$ used in the computation of $\Delta(u_i)$. During each iteration in Step 2 of **FEAS**, the retiming labels increase by at most 1, and before the update, the weight on each path $w(p_{i-1,i}) = 0$. After the update,

$$w_r(p_{i-1,i}) = r(u_i) - r(u_{i-1}) + w(p_{i-1,i}) = r(u_i) - r(u_{i-1}) \leq 1$$

Therefore, $\sum_{i=1}^{k} w_r(p_{i-1,i}) \leq k$. Since $P < d(p_{i-1,i})$ for each i, we have

$$P \cdot k < \sum_{i=1}^{k} d(p_{i-1,i})$$

$$\Rightarrow P \cdot \sum_{i=1}^{k} w_r(p_{i-1,i}) < \sum_{i=1}^{k} d(p_{i-1,i}) \tag{10.6}$$

As long as the clock period is chosen so that this condition is true, a feasible retiming will not be possible. Therefore, it is essential to choose a clock period for which this condition is violated; this provides an updated lower bound for the binary search, given by

$$P \geq \frac{\sum_{i=1}^{k} d(p_{i-1,i})}{\sum_{i=1}^{k} w_r(p_{i-1,i})} \tag{10.7}$$

The use of this bound to update the lower bound of the binary search was shown in [SR94] to provide tremendous improvements in the execution time of the algorithm.

10.4.2 The ASTRA approach

The relationship between clock skew and retiming.   The basis of the
ASTRA approach is the relationship between clock skew and retiming, as illustrated by the following example on a circuit with edge-triggered flip-flops. Let
us first consider the use of intentional clock skews for improving the circuit performance. Consider the circuit in Figure 10.6(a), which has unit delay inverters
and negative edge-triggered flip-flops (i.e., data is latched at the falling edge
of the clock). The delays of the first and the second combinational block are
3 units and 1, respectively. Therefore, the fastest allowable clock has a period
of 3 units. If an attempt is made to operate the circuit with a clock period of
2 units, then we have a long path violation at FF1, because the required data
arrival time at FF1 is 2 units (assuming zero setup time), but the data arrives
at 3 units, as shown in Figure 10.6(a). However, if a skew of +1 unit is applied
to the clock line to FF1, as shown in Figure 10.6(b), then the circuit can be
operated without any long path violation, even for a clock period of two units[2].

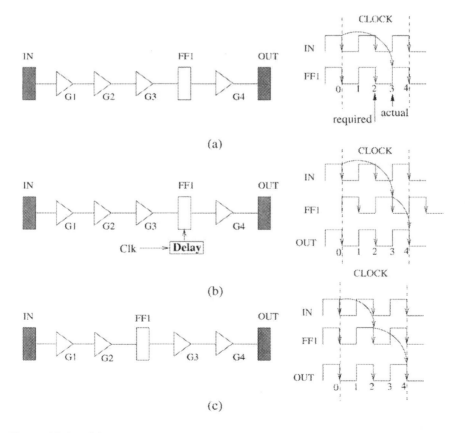

Figure 10.6.   (a) An example of a circuit with a long path violation. (b) Using clock skew
to correct the long path violation. (c) Correcting the long path violation using retiming.

It is easy to see that for the given circuit, the period can also be minimized to two units by retiming, if FF1 is moved to the left across the inverter G3, as shown in Figure 10.6(b).

In each case, one unit of time is borrowed by the first combinational block from the second; the manner in which cycle-borrowing occurs may either be via the vehicle of clock skew or via retiming. As a prelude to a more formal presentation of the relationship between clock skew and retiming, consider a flip-flop j in a circuit, as shown in Figure 10.7(a). As explained in Section 9.3.1, for every combinational path from a flip-flop i to j with delay $d(i,j)$, the following constraints must hold to ensure the absence of long path and short path violations, respectively:

$$x_i + \overline{d}(i,j) + T_{setup} \leq x_j + P$$
$$x_i + \underline{d}(i,j) \geq x_j + T_{hold}, \qquad (10.8)$$

where x_i and x_j are the skews at flip-flops i and j, respectively. Similar constraints can be written for every combinational path from flip-flop j to k with minimum delay $\underline{d}(j,k)$ and maximum delay $\overline{d}(j,k)$.

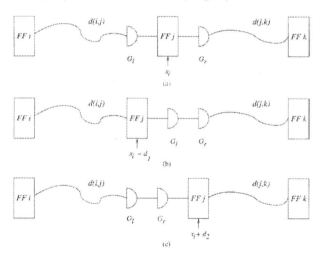

Figure 10.7.   Relationship between retiming and skew [DS95].

The relationship between retiming and skew can be captured in terms of Figure 10.7 [SD96]. For a circuit that operates at a clock period P, and satisfies the long path and short path delay constraints,

(a) retiming a flip-flop by moving it against the direction of signal propagation across a single-input, single-output gate of delay d_1 is equivalent to decreasing its skew by d_1.

(b) retiming a flip-flop by moving it in the direction of signal propagation across a single-output, single-output gate of delay d_2 is equivalent to increasing its skew by d_2.

This may be further generalized to multi-input multi-output combinational blocks is as follows:

(a) Retiming transformations may be used to move flip-flops from all of the n inputs of any combinational block to all of its outputs. The equivalent skew of the relocated flip-flop at output j, considering long path constraints only, is given by

$$x_j = \max_{1 \le i \le n} (x_i + \overline{d}(i, j)) \tag{10.9}$$

where the $x_i, 1 \le i \le n$, are the skews at the input flip-flops, x_j is the equivalent skew at output j, and $\overline{d}(i, j)$ is the worst-case delay of any path from i to j.

(b) Similarly, flip-flops may be moved from all of the m outputs of any combinational block to all of its inputs, and the equivalent skew at input k, considering long path constraints only, is given by

$$x_k = \min_{1 \le j \le m} (x_j - \overline{d}(k, j)) \tag{10.10}$$

where the $x_j, 1 \le j \le m$, are the skews at the input flip-flops, x_k is the equivalent skew at input k, and $\overline{d}(k, j)$ is the worst-case delay of any path from k to j.

In general, it is not possible to come up with an equivalent skew value that satisfies both long and short path constraints. For example, when we consider short path constraints, moving flip-flops from the input to the output requires that the new skew be

$$\min_{1 \le i \le n} (x_i + \underline{d}(i, j)) \tag{10.11}$$

which is incompatible with the requirement stated above except in the special case where all paths from i to j have the same delay. This is not an impediment here since the retiming problem as stated by Leiserson and Saxe in [LS91] considers long path constraints only.

Retiming may be thought of as a sequence of movements of flip-flops across gates. Starting from the final retimed circuit, where all of the skews are zero, and the long path constraints are met, this sequence of movements may be performed in reverse order. This procedure can be used to move all flip-flops back to their initial locations, using the above result to keep track of the altered clock skews at each flip-flop. The optimal retiming is equivalent to applying the set of skews thus obtained to the flip-flops in the circuit.

Note that the optimal clock period provided by the clock skew optimization procedure must, by definition, be no greater than the clock period for the set of clock skews thus obtained. Any differences arise due to the fact that clock skew optimization is a continuous optimization, while retiming is a discrete optimization. This argument leads to the following result:

Observation: The clock period obtained by an optimal retiming can be achieved via clock skew optimization. The clock period provided by the clock skew optimization procedure is less than or equal to that provided by the method of retiming.

Minimum period retiming by ASTRA. The relationship between skew and retiming motivates the following two-phase solution to the retiming problem:

Phase A: The clock skew optimization problem is solved to find the optimal value of the skew at each FF, with the objective of minimizing the clock period, or to satisfy a given (feasible) clock period, using the procedure described in Chapter 9.

Phase B: The skew solution is translated to retiming and some FF's are relocated across gates in an attempt to set the values of all skews to be as close to zero as possible. Flip-flops with positive skews are moved opposite to the direction of signal propagation (i.e., in the backward direction), and those with negative skews are moved in the direction of signal propagation (i.e., in the forward direction) to reduce the magnitude of their skews.

It can be proven [SD96] that at the end of this procedure, if all skews are set to zero, then the optimal clock period for this circuit is no more than $P_s + d_{max}$, where P_s is the optimal clock period found in Phase A, and d_{max} is the maximum delay of any gate in the circuit. This does not necessarily imply suboptimality for two reasons. Firstly, this is only an upper bound. Secondly, skew is a continuous optimization while retiming is discrete, and therefore, the clock period P_s achievable from the use of skews may not be attainable by retiming.

This method is extremely fast: experimental results on a mid-90s vintage workstations show that a 50,000 gate circuit can be retimed within two minutes using this procedure.

10.4.3   The ALAP and ASAP retimings

Retiming a circuit for a given target clock period is a special case of the minimum period retiming problem. Given a circuit and a clock period P, if the given clock schedule is feasible, then the method should return a retimed circuit that is correctly clocked, and if the clock schedule is not feasible, this should be indicated by the method. In the Leiserson-Saxe method, a specified-period retiming is obtained by simply running the algorithm **FEAS** with the target clock period. In the ASTRA approach for this problem, the binary search in Phase A is not performed, but for the target clock period P only, the constraint graph is constructed as before, and the Bellman-Ford algorithm is applied to obtain the set of required skews. If the Bellman-Ford algorithm detects a positive cycle, then the clock period is not feasible, and is reported as such; otherwise Phase B is performed.

A retiming for a given clock period is, in general, not unique, and different retimed circuits can be obtained, all of which satisfy the target clock period. This may be understood through the presence of slacks in the graph at the end of the Bellman-Ford procedure, which may be used to alter the skew values within a specified range without altering the feasibility. Out of the set of all possible specified-period retimings, two are of particular interest. The retiming that moves all registers as far as possible against the direction of signal propagation is called an "as soon as possible" (ASAP) retiming. Similarly, the retiming that moves all the registers as far as possible in the direction of signal propagation is referred to as the "as late as possible" (ALAP) retiming. Both ASAP and ALAP retiming assume that no register is moved across the host node.

In this section we will concentrate on explaining the ASAP and ALAP retimings in the context of edge-triggered circuits. However, the reader should remain aware that the same basic principle applies to level-clocked retimings, even though we do not address the issue in detail here.

The ASAP and the ALAP locations can be seen as the extreme locations of the locations of the registers in the circuit for the specified clock period, and are *unique*. Their importance is in the fact that they may be utilized to make the procedure of minimum area retiming more efficient, as explained in Section 10.5.2.

For the example circuit in Figure 10.8(a), with unit delay inverters and edge-triggered FF's, the ASAP and ALAP retimings for a target clock period of 3 units are shown in Figure 10.8(b) and Figure 10.8(c), respectively.

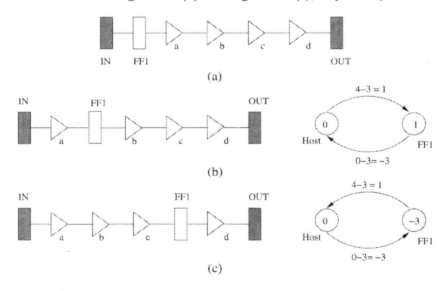

Figure 10.8. (a) An example circuit. (b) The ASAP retiming of the circuit, and the corresponding the ASAP constraint graph. (c) The ALAP retiming of the circuit, and the ALAP constraint graph [MS97a].

In ASAP retiming, the objective is to move the FF's as far as possible in the backward direction. Since Phase B of ASTRA moves FF's with positive skews in the backward direction, an ASAP retiming would aim to obtain the maximum possible skew value for each register in Phase A. Similarly, it can be argued that FF's with negative skews would be at their most backward location if the skews were set to be at their maximum (least negative) value. Therefore, in either case, an ASAP retiming corresponds to the maximum allowable skew values; symmetrically, an ALAP retiming corresponds to the minimum possible (least positive and most negative) skew values. The ASAP skews are obtained by running the Bellman-Ford algorithm on the transpose [CLR90] of the original constraint graph, i.e., a graph with the same vertex set as the original graph, but with the edge directions reversed. In this transpose graph, an edge from u to v exists if there is a combinational path from FF v to FF u; the weight of this edge is $\bar{d}(v, u) - P$. The initial skew values for the Bellman-Ford algorithm are set to $-\infty$ for all FF's, except the host which is initialized to 0. Since all the edge directions have been reversed, the longest path values obtained for each node in the graph must be made to undergo a sign reversal to obtain the correct skew values for the corresponding FF's. The skew values so obtained are the maximum possible for the specified clock period, and no skew can be increased further without violating the clock period. For example, the transposed Phase A constraint graph for ASAP retiming of the circuit in Figure 10.8(a), is shown in Figure 10.8(b). The longest path to the node corresponding to FF1 is $+1$ unit, and hence the corresponding skew on FF1 of -1 unit is obtained by reversing the sign.

Performing Phase B with these skew values results in the ASAP retiming. For ASAP locations, the available slack is used to avoid moving an FF in the direction of signal flow. The procedure for finding the ASAP (and ALAP) retiming proceeds along the same lines as in Section 10.4.2, with a few variations described below.

To obtain the ASAP locations for the retimed FF's, it is necessary to push the FF's as far as possible in a backward direction. Therefore, each FF with positive skew is moved as far as possible in the backward direction, and each FF with negative skew is moved as little as possible in the forward direction. Therefore,

(1) for an FF with positive skew x that is being moved across a single-fanout gate p in the backward direction, the skew value after the relocation at input i of p is set to $x - \text{delay}(p)$. If this value is nonpositive, then the ASAP location has been found. For gates with multiple fanouts, $x = \min_{\text{all outputs}}(x_i)$, where x_i is the skew of the FF at the ith output, as shown in Figure 10.9(a).

(2) for an FF with negative skew x that is being moved across a single-fanin gate p in the direction of signal propagation, the skew value after the relocation at output i of p is set to $x + \text{delay}(p) + \text{slack}(i)$, where $\text{slack}(i)$ is the slack associated with the output i. This slack is defined as the amount by which the delay at output i may be increased before it becomes the critical output of p; by definition, the critical output has zero slack. If the new skew is nonnegative, then the ASAP location has been found. For gates with multiple fanins, $x = \max_{\text{all inputs}}(x_i)$, where x_i is the effective skew of the FF at the ith output, as shown in Figure 10.9(b).

Applying this procedure to circuit in Figure 10.8(a) requires us to move FF1, which has a skew of -1 unit, across the unit delay inverter a to the location shown in Figure 10.8(b). At this point, the skew on FF1 is down to zero, and hence the ASAP location has been reached. As is shown in this example if the initial circuit does not satisfy the target clock period, then ASAP retiming may require FF's to move forward, and these forward moves will be required by *all* retimings satisfying the target clock period.

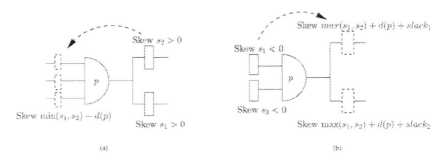

Figure 10.9.   Effective skews at FF's after ASAP retiming across a gate.

Since the objective in ALAP retiming is to move the FF's as far as possible in the forward direction, Phase A of ASTRA is modified to yield the minimum (most negative and least positive) skew values. All skew values are initialized to $-\infty$, except the host which is set to zero, before applying the Bellman-Ford algorithm to find the longest path in the constraint graph. In Phase B of ALAP retiming, each FF is moved as far as possible in the forward direction, using slacks if any to minimize any FF movements in the backward direction.

The Phase A constraint graph for ALAP retiming of the circuit in Figure 10.8(a) is shown in Figure 10.8(c). The longest path to the node corresponding to FF1 is +3 units, and hence the skew on FF1 is 3 units. Since FF's with positive skews are moved as far forward as possible for ALAP retiming, FF1 moves across gates a, b, and c until its skew is brought down to zero and the ALAP location has been achieved.

10.4.4   Continuous retiming

A related technique to the ASTRA algorithm is the procedure of continuous retiming (c-retiming) [Pan97]. Unlike retiming, this method assigns a real number to each gate, representing the fractional number of FF's to be moved across the gate. The motivation for c-retiming is its potential use in studying synthesis and optimization problems that involve circuit modifications in conjunction with retiming, and an application of c-retiming to the tree mapping problem [De 94] is presented in [Pan97]. Rounding off this fractional number of FF's will result in a conventional retiming, with the resulting clock period guaranteed to be within the largest gate delay of the optimal clock period. This provides an efficient way for minimum period retiming, since c-retiming can be solved as a single-source longest path problem. In c-retiming, the edge weights ($w_s(e)$) are also real numbers and denote the number of fractional FF's present on an edge e. The clock period of the c-retimed circuit is P if the final continuous retimed weight on each edge is no smaller than $\frac{d(v)}{P}$.

10.5   MINIMUM AREA RETIMING OF EDGE-TRIGGERED CIRCUITS

As shown in Section 10.1, retiming can reduce the clock period of a circuit. However, in doing so, it is possible that it may greatly increase the number of registers. In fact, given any clock period for a general circuit, there will, in general, be not one, but several retiming solutions that satisfy the clock period; these differ in the manner in which they utilize the slack on noncritical paths. The ASAP and ALAP retimings from Section 10.4.3 are just two of these possible retimings, and each of these differs in terms of the number of registers that it employs.

The objective of the minimum area retiming problem is to find one of these retiming solutions at a given clock period, chosen in such a way that it has the minimum number of registers over all feasible retimings satisfying the period. This is a practical and meaningful objective, since minimizing the number of registers would minimize the area of the circuit, as the retiming transformation leaves the combinational logic untouched.

We will first overview the Leiserson-Saxe framework for minimum area retiming. Next, an efficient method for reducing the complexity of this method, using bounds on the retiming variables derived from the equivalence between clock skew and retiming, is presented. Finally, the extension of this approach to level-clocked circuits is described.

10.5.1   The Leiserson-Saxe approach

A basic formulation.   A mathematical programming formulation of the minimum area retiming problem was presented in [LS91], and is reproduced here. Let the total number of registers in a circuit G be given by $S(G) = \sum_{e \in E} w(e)$. The reader is referred back to Section 10.3 for the basic concepts and terminology used in the representation of a circuit in terms of a retiming graph. Using this notation, the total number of registers in a circuit after retiming, $S(G_r)$, can be calculated as follows:

$$S(G_r) \;=\; \sum_{e \in E} w_r(e) \tag{10.12}$$

$$=\; \sum_{e:u \to v \ \in E} w(e) + r(v) - r(u) \tag{10.13}$$

$$=\; S(G) + \sum_{v \in V} r(v)(|FI(v)| - |FO(v)|) \tag{10.14}$$

where $FI(v)$ and $FO(v)$ represent the fanin and fanout sets of the gate v. Since $S(G)$ is a constant, the minimum area retiming problem for a target period P

can be formulated as the following linear program (LP):

$$\text{minimize} \sum_{v \in V} [(|FI(v)| - |FO(v)|) \cdot r(v)] \qquad (10.15)$$

$$\text{subject to} \qquad r(u) - r(v) \leq w(e_{uv}) \qquad \forall e_{uv} \in E$$

$$r(u) - r(v) \leq W(u, v) - 1 \qquad \forall D(u, v) > P$$

$$-\infty \leq r(u) \leq \infty \qquad \forall u \in V$$

The significance of the objective function and the constraints is as follows.

- The objective function represents the number of additional registers added to the retimed circuit, with reference to the original circuit.

- The first constraint ensures that the weight e_{uv} of each edge (i.e., the number of registers between the output of gate u and the input of gate v) after retiming is nonnegative. We will refer to these constraints as *circuit constraints*.

- The second constraint ensures that after retiming, each path whose delay is larger than the clock period has at least one register on it. These constraints, being dependent on the clock period, are often referred to as *period constraints*.

This problem formulation is in the form of a dual of a minimum-cost network flow problem. Hence, the LP can be solved efficiently by solving this dual [BJS77, AMO93].

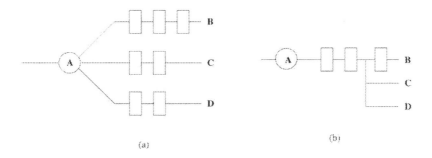

Figure 10.10.   Unconditional register sharing at multiple fanouts.

A more accurate area model using mirror vertices.   The cost function in the LP (10.15) assumes that each FF has exactly one fanout. However, in practice an FF can have multiple fanouts, allowing the FF's on different fanout edges of a gate to be shared. This sharing must be taken into account for an accurate area model.

As an example, consider gate A in Figure 10.10 with three fanouts, B, C, and D, having three, two and two FF's, respectively. The LP in (10.15) will model the total number of FF's as seven as shown in Figure 10.10(a). However, the FF's can be merged or shared as shown in Figure 10.10(b), resulting in a total cost of only three FF's.

To model the maximal FF sharing, the work in [LS91] introduces a mirror vertex m_i for each gate i that has more than one fanout, as shown in Figure 10.11, the details of which can be found in [Sax85]. Each edge e_{ij}, in addition to having a weight $w(e_{ij})$, now also has a width $\beta(e_{ij})$. In Figure 10.11, the edge weights are shown above the edges while the edge widths are shown below the edges. Consider a gate u with k fanouts to gates v_j, $j = 1, \cdots, k$. To model the maximum sharing of FF's, an extra edge is added from each fanout gate, v_j, to the mirror vertex, m_u, with a weight of $w(e_{v_j m_u}) = w(\max_u) - w(e_{uv_j})$. Here, $w(\max_u) = \max_{\forall i \in FO(u)}(w(e_{ui}))$ is the maximum weight on any fanout edge of gate u. Each of the edges from the gate i to its fanouts j, and from the fanouts to the mirror vertex has a width of $1/k$, i.e.,

$$\beta(e_{uv_j}) = 1/k \text{ and } \beta(e_{v_j m_u}) = 1/k \quad \text{for } j = 1, \cdots, k.$$

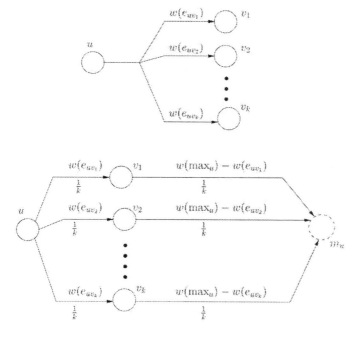

Figure 10.11.   Model for maximum register sharing at multiple fanouts [LS91].

The original LP in (10.15) is modified to include the effect of register sharing as follows:

$$\min \sum_{v \in (V \cup M)} \left[\left(\sum_{\forall j \in FI(v)} \beta(e_{jv}) - \sum_{\forall j \in FO(v)} \beta(e_{vj}) \right) \cdot r(v) \right]$$

subject to
$$r(u) - r(v) \le w(e_{uv}) \qquad \forall e_{uv} \in E \qquad (10.16)$$
$$r(u) - r(v) \le W(u,v) - 1 \qquad \forall D(u,v) > P$$
$$r(j) - r(m_i) \le w(\max_i) - w(e_{jm_i}) \qquad \forall (m_i) \in M \text{ and } \forall j \in FO(i)$$
$$-\infty \le r(u) \le \infty \qquad \forall u \in (V \cup M)$$

where $M = \{m_v | v \in V \text{ and } |FO(v)| > 1\}$ is the set of all the mirror vertices, and additional constraints due to the mirror vertices are called the mirror constraints. For simplicity, we can rewrite the above LP as follows

$$\min \sum_{v \in (V \cup M)} \left[\left(\sum_{\forall j \in FI(v)} \beta(e_{jv}) - \sum_{\forall j \in FO(v)} \beta(e_{vj}) \right) \cdot r(v) \right]$$

$$\text{subject to} \quad r(u) - r(v) \le c_{uv} \quad \forall (u,v) \in C \qquad (10.17)$$
$$-\infty \le r(u) \le \infty \quad \forall u \in (V \cup M)$$

where $C = C_p \cup C_c \cup C_m$ is the constraint set of the LP in (10.16), and includes the period constraint set (C_p), the circuit constraint set (C_c) and the mirror constraint set (C_m). A constraint (i, j) in the constraint set C is of the form

$$r(i) - r(j) \le c_{ij} \forall (i,j) \in C$$

where

$$
\begin{array}{llll}
c_{ij} &=& w(e_{ij}) & \forall (i,j) \in C_c, \text{ i.e., } e_{ij} \in E \\
c_{ij} &=& W(i,j) - 1 & \forall (i,j) \in C_p, \text{ i.e., } D(i,j) > P \qquad (10.18) \\
c_{ij} &=& w(\max_i) - w(e_{jm_i}) & \forall (i,j) \in C_m, \text{ i.e., } m_i \in M, \forall j \in FO(i)
\end{array}
$$

The objective function of the LP in (10.17) now denotes the increase in the number of FF's assuming maximal sharing of FF's at the output of all gates. The weights on all paths from gate u to its mirror vertex m_u are the same before retiming, i.e., $w(e_{uv_i}) + w(e_{v_i m_u}) = w(\max_i), 1 \le i \le k$, and therefore, the weights on all paths from gate u to its mirror vertex m_u must be equal after retiming. Since the mirror vertex m_u is a sink in the graph, the register count on one of the edges from the fanout nodes to m_u will be zero, i.e., $\exists i \ |w(e_{v_i,m_u}) = 0$. Thus, the weight on all paths from gate u to mirror vertex m_u after retiming will be $w_r(\max_u) = \max_{\forall j \in FO(u)}(w_r(e_{uj}))$, since all of the retimed edge weights $w_r(e_{jm_u}) \ge 0 \forall j \in FO(u)$. As there are k paths, each with width $1/k$, the total cost of all paths will be $w_r(\max_u)$ as desired. Like the LP in (10.15), the LP in (10.17) is also the dual of a minimum cost network flow problem.

An alternative view of this model is as follows. The change in cost function due to adding or removing latencies from the fanout junction of gate u is modeled by two retiming variables: one for the gate, $r(u)$ and other for the mirror vertex, $r(m_u)$. Any change in the cost function due to FF's moving across the multi-fanout gate itself are modeled by $r(u)$, while any change due to FF motion across its fanout gates v_i, $1 \leq i \leq k$ is modeled by the mirror variable $r(m_u)$.

The change in the number of FF's in the circuit, under maximal sharing obtained by retiming a gate u by one unit can be calculated as follows. The decrease in the cost function obtained by removing an FF from each of the fanouts of a gate is one unit, even for multiple fanout gates since the FF's on all the fanouts were shared. The increase in the cost function from adding an FF to all the inputs of a gate u is equal to the number of fanins of u that have only one fanout, since any FF added to a fanin j of gate u that has more than one fanout ($|FO(j)| > 1$) is already modeled by the mirror variable of that fanin gate m_j. Thus, the cost contribution of any single fanout gate u is given by $(|FI'(u)| - 1) \cdot r(u)$, while that of a multi-fanout gate is given by $(|FI'(u)| - 1) \cdot r(u) + r(m_u)$, where $FI'(u)$ is the set of fanins that have only a single output, i.e., $FI'(u) = \{v | v \in FI(u)$ and $|FO(v)| = 1\}$. Therefore, the cost function may be written as the summation of these terms over all gates.

Efficient implementation. Although the LP in (10.15) can be efficiently solved by solving the dual, the computation of W and D matrices requires $O(|V|^3)$ time and $O(|V|^2)$ memory. Further, the number of period edges required can be very large, destroying the sparsity of the circuit graph. Because of these reasons, the minimum area retiming problem cannot be solved in reasonable time using the method from [LS91]. Shenoy and Rudell in [SR94] presented an efficient implementation, where they proposed a constraint generation method requiring $O(|V|^2|E|log|V|)$ time but only $O(|V|)$ memory, and a technique to prune the number of period constraints.

The algorithm uses a combination of the Dijkstra's algorithm and the Bellman-Ford algorithm. It works by generating one row, say the (s^{th}) row, of the W and the D matrix at a time. An ordered pair $(w(e_{ij}), -d(i))$, denoted by (a_i, b_i), is associated with each edge e_{ij} and is used to compute the shortest distance from a source vertex s. A heap is maintained for each distinct value of a_i and is indexed by this value. Until all heaps are empty, the node u at the top of the minimum index heap is extracted using the function pop-min(heap index). The fanouts of u are then added to the appropriate heaps if their a_u or b_u values are updated by Bellman-Ford relaxation. At the end of this procedure $D(s, u) = -b_u$ and $W(s, u) = a_u$.

Note that to satisfy a clock period, P, the only requirement is to ensure that each path with delay greater than P has at least one FF on it. The number of FF's on any path is monotonic with respect to the path length since negative edge weights are not allowed. Due to this monotonicity of edge weights, if at least one FF is placed on a sub-path, then it is guaranteed that at least one FF will exist on all paths containing this sub-path. Adding a period constraint from s to u is one way to ensure at least one FF on all paths from s to u. The idea is to add a period edge to only the vertex v, reachable from s, that satisfies the following:

$$D(s,v) > P \text{ and } D(s,u) \leq P \; \forall u \text{ on } s \rightsquigarrow FI(v) \qquad (10.19)$$

where $s \rightsquigarrow FI(v)$ is a path from s to a fanin of v. Thus, if the period constraint is added, the fanouts of u need not be relaxed. The pseudocode for constraint generation is as follows:

```
Algorithm CONSTRAINT_GENERATION
{
  P = target clock period;
  S_k = the k^th heap;
  ∀s ∈ V {
    s = current vertex;
    ∀v ∈ V, a_v = ∞ and b_v = 0;
    S_0 = {s}; a_s = 0,; b_s = -d(s);
    k = current register weight;
    do {
        k = min{p | S_p ≠ ∅};
        u = pop-min(S_k) ;
        if (-b_u > P)
            add a period edge c(s,u) with weight a_u - 1
        else {
            ∀v ∈ FO(u) {
                if((a_v, b_v) > (a_u + w(e_u,v), b_v - d(v)))
                heap-insert(S_{a(u)+w(e_u,v)}, v) ;
            }
        }
    } while(∃ p | S_p ≠ ∅)
  }
}
```

The work in [vVA+94], published a few months before [SR94], used the same idea and referred to it as "clock-period limited labeling." It was demonstrated that the use of this method caused a significant speedup in constraint generation. A second method presented in the work was termed "relevant path labeling." Starting from a source node u, this technique labels each vertex v with the maximum value of the label $l(u,v) = \frac{D(p(u \rightsquigarrow v))}{P} - W(p(u \rightsquigarrow v))$. The rationale behind this approach is that for every path originating at u, it must be true that $\frac{D(p(u \rightsquigarrow v))}{P} \leq W(u \rightsquigarrow v)$, and therefore, the largest value of $l(u,v)$ must be nonpositive. However, unlike the method of clock-period limited labeling, this must traverse all vertices, and while it generates a smaller number of constraints, the amount of time required to generate these constraints was reported to be extremely large.

10.5.2   The Minaret algorithm

Although the techniques of the previous section make minimum area retiming efficient, they cannot handle large circuits with tens of thousands of gates. In [MS97a, MS98b], an amalgamation of the Leiserson-Saxe approach and the

ASTRA approach is used for efficient minimum area retiming. By utilizing the merits of both approaches, an efficient algorithm for constrained minimum area retiming was developed. This algorithm, called MINArea RETiming (Minaret), is capable of handling very large circuits in very reasonable runtime. The basic idea of the approach is to use the ASTRA approach to find tight bounds on the retiming variables. These bounds help reduce both the number of variables and the number of constraints in the problem without any loss in accuracy. By spending a small amount of additional CPU time on the ASTRA runs, this method leads to significant reductions in the total execution time of the minimum area retiming problem. The reduction in the problem size also reduces the memory requirements, thus enabling retiming of large circuits. On a mid-to-late 90s vintage computer, this approach has been shown to solve problems with more than 57,000 variables and 3.6 million constraints in about 2.5 minutes.

The approach in Minaret is to find tight bounds on the r variables, and to use these bounds to avoid generating redundant constraints. By appropriate application of these bounds, not only is the constraint set pruned but the number of variables is also reduced. In this way, the size of the LP is reduced enabling it to be solved more efficiently. The reduced constraints can be generated efficiently by using these bounds. Note that the exactness of the solution is not sacrificed in doing so, since none of the essential constraints are removed.

We will now show the relation between the Leiserson-Saxe approach and the ASTRA approach, and how a modified version of ASTRA can be used to derive bounds on the r variables in the Leiserson-Saxe method. Next, we show how these bounds can be used to prune the number of constraints in minimum area Leiserson-Saxe retiming. Finally, we present an example to illustrate the method.

The concept of restricted mobility and bounds for the r variables.
For the circuit in Figure 10.12, to achieve the minimum clock period of 4.0 units, one must move one copy of FF B to the output of gate G4. The possible locations for FF's along the other path to FF C are at the input to gate G8, or at the output of gate G8, or the inputs of gates (G9,G10) or the outputs of gates (G9,G10); no other locations are permissible

Therefore, it can be seen that the FF's cannot be sent to just any location in the circuit; rather, there is a restricted range of locations into which each FF may be moved, and the mobility of each FF is restricted. This restricted mobility may be used to reduce the search space, and hence the number of constraints.

The concept of restricted mobility is related to the ASAP and ALAP retimings presented in Section 10.4.3 in that the ASAP and ALAP retiming solutions define the boundaries of the region into which an FF can move during retiming while satisfying the target clock period. For example, Figures 10.12(a) and (b) are the ASAP and ALAP retiming for the given clock period of 4.0 units and the FF can move only in the region defined by these locations. These ASAP and ALAP retiming solutions can be used to obtain bounds on the retiming

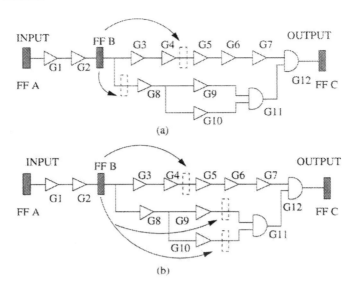

Figure 10.12. A set of possible FF locations after retiming for a clock period of 4 units (all gates are assumed to have unit delays) [MS97a].

variables of the Leiserson-Saxe approach, r, associated with the gates in the circuit as illustrated by the following example.

Example: For the circuit in Figure 10.12, the locations for the FF's in the retimed circuit corresponding to the ASAP and ALAP retiming solutions are shown in Figure 10.12(a) and (b), respectively. This implies that during retiming, no latency will move across gates G1, G2, G5, G6, G7, G11 and G12; one latency each will move from the input to the output of gates G3 and G4, and either no latency or one latency will move from the input to the output of gates G8, G9 and G10. Referring to Section 10.3 for the definition of the r variables, this implies that one may set the following bounds on the r variables.

(1) $r(u) = 0$ for $u \in \{G1, G2, G5, G6, G7, G11, G12\}$
(2) $r(u) = -1$ for $u \in \{G3, G4\}$, and
(3) $-1 \leq r(u) \leq 0$ for $u \in \{G8, G9, G10\}$.   □

While moving the FF's in Phase B of ASAP and ALAP retimings, subject to the specified clock period P, the number of FF's that traverse each gate is counted; this count is the upper and lower bound, respectively, on the r variables for each gate. An FF moving from the inputs to the output of a gate decrements the count by one, while one moving from the output to the inputs increments it by one. For the ASAP case, FF's are moved as far as possible in the backward direction. In other words, we relocate the largest number of latencies possible from the output to the inputs of a gate. By the definition of the r variables, this gives an upper bound on r for the gates. Similarly, the ALAP retiming relocates the largest number of latencies that can move from the inputs of a gate towards its output, and this provides a lower bound on the

r values for the gates in the circuit. The bounds on the r variable corresponding to each gate y are of the form.

$$L_y \leq r(y) \leq U_y \qquad (10.20)$$

We will refer to L_y as the lower bound for gate y and to U_y as the upper bound of gate y. Like the ASAP and ALAP retimings, these bounds are with reference to a fixed host vertex, i.e., $L_{\text{host}} = U_{\text{host}} = 0$. If $U_u = L_u = k_u$, we say that gate u is *fixed* or immobile, since $r(y) = k_y$ is not really a variable any more. On the other hand, if $U_y \neq L_y$, we say that gate y is *flexible* or mobile. Thus, we can reduce the variable set V of the Leiserson-Saxe model to $V' \subseteq V$, the variable set of Minaret where

$$V' = \{v \in V | U_v \neq L_v\} \qquad (10.21)$$

Bounds on the mirror vertices, introduced to model the maximal latch sharing can be obtained directly from the bounds on fanout gates. The mirror variable set M is also reduced to $M' \subseteq M$ the mirror variable set of Minaret where

$$M' = \{m \in M | U_m \neq L_m\} \qquad (10.22)$$

The bounds on the r value of a mirror vertex m_i of gate i in Figure 10.11 can easily be derived from the bounds on the fanout gates and are given by

$$U_{m_i} = \max_{\forall j \in FO(i)} (U_j + w(e_{ij})) - w(max_i)$$
$$L_{m_i} = \max_{\forall j \in FO(i)} (L_j + w(e_{ij})) - w(max_i)$$

For a proof, the reader is referred to [MS98b].

Eliminating unnecessary constraints. We now illustrate how the addition of bounds (derived previously) to the LP of (10.17) in Section 10.5.1 may be used to reduce the constraint set by dropping redundant constraints. It can be seen from the bounds on $r(i)$ and $r(j)$ in relation (10.20) that $r(i) - r(j) \leq U_i - L_j$. Therefore, if $U_i - L_j \leq c_{ij}$ then $r(i) - r(j) \leq c_{ij}$ is also true, and the constraint (i, j) can be dropped. Thus, the Leiserson-Saxe constraint set C can be reduced to the Minaret constraint set $C' \subseteq C$ where

$$C' = \{(i,j) \in C | \ U_i - L_j > c_{ij}\} \qquad (10.23)$$

Notice that constraints associated with fixed or immobile gates can be treated as bounds and need not be included in C'. Like the Leiserson-Saxe constraints, the Minaret constraints also consists of circuit, period and mirror constraints, i.e., $C' = C'_c \cup C'_p \cup C'_m$, where C'_c is the reduced circuit constraint set, C'_p is the reduced period constraint set, and C'_m is the reduced mirror constraint set.

The reduced linear program. We use the relations (10.21), (10.22) and (10.23) to reduce the LP in (10.17) to the following LP in Minaret

$$\min \sum_{v\in\{V'\cup M'\}} \left[\left(\sum_{\forall j\in FI(v)} \beta(e_{jv}) - \sum_{\forall j\in FO(v)} \beta(e_{vj})\right) \cdot r(v)\right] \qquad (10.24)$$

$$\text{subject to} \quad r(u) - r(v) \le c_{uv} \;\; \forall(u,v) \in C'$$
$$L_u \le r(u) \le U_u \;\; \forall u \in (V' \cup M')$$

Details of how the linear program is generated and solved are presented in [MS98b].

An Example: We now illustrate this method and show how the number of constraints can be reduced using our approach.

Consider the circuit example shown in Figure 10.13. As in the previous examples, we make the assumption that the gates have unit delays. We consider two possible clock periods of 2 units and 3 units in this example.

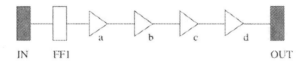

IN   FF1   OUT

Figure 10.13.   An example illustrating the retiming algorithm [MS97a].

When $P = 2$ units

For a clock period of two units, the list of constraints generated by the approach in [SR94] is listed below.

$$
\begin{array}{lll}
\underline{\text{Circuit constraints}} & r(h) - r(a) & \le 1 \\
& r(a) - r(b) & \le 0 \\
& r(b) - r(c) & \le 0 \\
& r(c) - r(d) & \le 0 \\
& r(d) - r(h) & \le 0 \\
\underline{\text{Period constraints}} & r(h) - r(c) & \le 0 \\
& r(a) - r(c) & \le -1 \\
& r(b) - r(d) & \le -1 \\
\end{array}
$$

Note that

(a) the delay associated with the host node is zero, and

(b) the value of $r(h)$ is set to zero as a reference, so that it is not really a variable.

Therefore, this is a problem with <u>four</u> variables and <u>eight</u> linear constraints (of which three act as simple bounds).

In our approach, for a clock period of 2, we first find the bounding skews. The FF's at the input and output may not be moved, and therefore, the only movable FF is FF1, which is assigned a skew of -2 units. The correctness of this skew value is easy to verify since the only feasible location of FF1 under $P = 2$ is two delay units to the right of its current location. Therefore, we find that by using the concept of restricted mobility,

$$-1 \le r(a) \le -1 \;\; \Rightarrow r(a) = -1$$
$$-1 \le r(b) \le -1 \;\; \Rightarrow r(b) = -1$$
$$0 \le r(c) \le 0 \;\;\;\; \Rightarrow r(c) = 0$$
$$0 \le r(d) \le 0 \;\;\;\; \Rightarrow r(d) = 0$$

Since all nodes are fixed, and all the constraints can be dropped, all of the constraints and variables have been eliminated!

When $P = 3$ units

With the clock period is set to 3 units, the list of constraints is

$$\begin{array}{lll} \underline{\text{Circuit constraints}} & r(h) - r(a) & \le 1 \\ & r(a) - r(b) & \le 0 \\ & r(b) - r(c) & \le 0 \\ & r(c) - r(d) & \le 0 \\ & r(d) - r(h) & \le 0 \\ \underline{\text{Period constraints}} & r(h) - r(d) & \le 0 \\ & r(a) - r(d) & \le -1, \end{array}$$

As before, $r(h) = 0$ is set as a reference, giving a problem with <u>four</u> variables (as before) and <u>seven</u> linear constraints (of which three act as simple bounds).

Under our approach, the relocated FF can reside either at the input of gate b, the output of gate b, or the output of gate c. Therefore, we have

$$-1 \le r(a) \le -1 \;\; \Rightarrow r(a) = -1$$
$$-1 \le r(b) \le 0$$
$$-1 \le r(c) \le 0$$
$$0 \le r(d) \le 0 \;\;\;\; \Rightarrow r(d) = 0$$

Using these bounds we drop all constraints but

$$r(b) - r(c) \le 0$$

Therefore, we have reduced the problem complexity to <u>two</u> variables, each with fixed upper and lower bounds and <u>one</u> linear constraint. (Note that upper/lower bound constraints are typically much easier to handle in LP's than general linear constraints; in fact, in many cases, upper and lower bounds are actually helpful in solving the LP.)   □

Generating the linear program. Using the alternative description of the maximal FF sharing in Section 10.5.1, the objective function coefficients are obtained by inspection of the circuit, without explicitly adding the mirror vertices. The circuit and the mirror constraints in C' are obtained from direct inspection of the circuit graph using relation (10.23). Because the bounds on the mirror vertices can also be obtained directly from the bounds on the gate vertices, we do not need to explicitly add the mirror vertices to the circuit graph. Since every multi-fanout gate has a mirror vertex, this gives us important savings in terms of the space and time requirements. We now describe how to obtain the period constraints in C'.

We take advantage of the bounds obtained in Section 10.5.2 to modify the method from [SR94] to run faster, generating only the reduced constraint set C'.

As noted earlier, due to the monotonicity of edge weights, if we ensure at least one FF on any sub-path, we are assured of having at least one FF on all paths that contain this sub-path. Therefore, if the bounds on the r variables guarantee us at least one FF on any sub-path, we need not process any path containing this sub-path. We use this observation in addition to relation (10.19) to reduce the number of period constraints.

At the end of the ASTRA run for obtaining the lower bounds, all FF's are in their ALAP locations. If the delay of all the gates is not the same, it is possible that retimed circuit obtained by ASTRA with FF's in the ALAP locations may have some purely combinational paths with delays that are greater than the target clock period P. However, in practice, most of the other paths satisfy the target clock period. We will use this observation to further speed up the constraint generation process.

Consider a fixed gate a in the circuit at the end of the ALAP run. If none of the combinational paths starting at this gate violate the clock period, we have $W_{ALAP}(a, i) \geq 1$ if $D(a, i) > P$ $\forall i$. Since $W_{ALAP}(a, i) = W(a, i) + L_i - L_a$ we have $L_a - L_i \leq W(a, i) - 1$, or $L_a - L_i \leq c_{a,i}$. Since gate a is fixed $U_a = L_a$, we obtain $U_a - L_i \leq c_{a,i}$ $\forall i \in V$, which is guaranteed to be true, and hence, all constraints starting from fixed gate a are redundant, and we do not need to generate them. Thus, we must generate period constraints only from those fixed gates which have at least one purely combinational path starting from it with delay more than the clock period. Let us call this set V''.

The method from [SR94] presented in Section 10.11 is modified to take advantage of the bounds on the r variables to generate the reduced constraint set C' efficiently. The pseudocode is presented below with the modifications shown in boldface.

```
Algorithm CONSTRAINT_GENERATION_WITH_BOUNDS
{
  P = target clock period;
  L_i ≤ r(i) ≤ U_i  ∀i ∈ V;
  S_k = the k^{th} heap;
  L_{min} = min(L_i)  ∀i ∈ V;
  ∀s ∈ V' ∪ V''  {
      s = current vertex;
      ∀v ∈ V, a_v = ∞ and b_v = 0;
      S_0 = {s}, a_s = 0,  and  b_s = -d(s);
      k = current register weight;
      do {
          k = min{p | S_p ≠ ∅};
          if (k ≥ (U_s - L_{min} + 1)) break;
          u = pop-min(S_k) ;
          if ((U_s - L_u) ≤ k - 1) continue;
          if (-b_u > P)
          add a period edge c(s,u) with weight a_u - 1
          else {
                  ∀v ∈ FO(u) {
                      if ((k - U_s + L_v) < 1)   {
                          if((a_v, b_v) > (a_u + s_{u,v}, b_v - d(v)))
                          heap-insert(S_{a(u)+s_{u,v}}, v);
                      }
                  }
          }
      } while(∃ p | S_p ≠ ∅)
  }
}
```

Solving the linear program. Like (10.17), the LP in (10.24) is also a dual of a minimum cost network flow problem. We found that it could be solved very efficiently using the network simplex algorithm from [BJS77]. The network simplex method is a graph based adaptation of the LP simplex method that exploits the network structure to achieve very good efficiency. The upper and lower bounds on the r variables provide an initial feasible spanning tree. This tree has two levels only, with the host node as the root and all other nodes as leaves. To prevent cycling we construct the initial basis to be strongly feasible by using the appropriate bound (upper or lower) to connect a node to the root (host node). It is easy to maintain strongly feasible trees during the simplex operations, and details are given in [BJS77].

Using the first eligible arc pivot rule with a wraparound arc list from [AMO93] (page 417) gives significant improvements in the run time. The dual variables (r variables) are directly available from the min cost flow solution.

10.6   MINIMUM PERIOD RETIMING OF LEVEL-CLOCKED CIRCUITS

For edge-triggered circuits, the delays through all combinational logic paths must be less than the clock period, with allowances for the setup time. Therefore, timing constraints need only be enforced between FF's connected by a purely combinational path. For level-clocked circuits, the delay through a combinational logic path may be longer than one clock cycle, as long as it is compensated by shorter path delays in subsequent cycles. To ensure that the extra delay is compensated, timing constraints must be enforced from each latch to every other latch reachable from it, possibly through a path that traverses multiple latches. This greatly increases the complexity of the problem for practical circuits, as was illustrated by the example of an N-stage acyclic pipeline in Section 7.3.1.

10.6.1   Variation of the critical path with the clock period

As in the case of edge-triggered circuits, traditional methods [LE94, PR93b] solve the minimum period retiming problem for level-clocked circuits by performing a binary search over all possible clock periods. At each step of this binary search, the feasibility of achieving the clock period by retiming is checked by solving a single source shortest path problem, using the Bellman-Ford algorithm on a constraint graph. For a circuit with $|G|$ gates, this constraint graph consists of $|G|$ vertices and edges between every pair of vertices, constructed by solving an all-pairs shortest path problem on the original circuit graph. This graph must be reconstructed for every binary search point, because will be shown shortly, unlike in the case of edge-triggered circuits, critical paths in level-clocked circuits can be different for different clock periods. Therefore, the methods in [PR93b, LE94] have $O(|G|^2)$ memory requirements and high (although polynomial) time complexity.

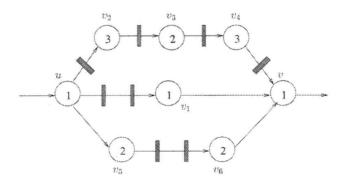

Figure 10.14.   An example illustrating the change in the critical path with the clock period [LE94].

In an edge-triggered circuit, the critical path is independent of the clock period, and simply corresponds to the output with the largest delay, except in the uninteresting case when the clock period is satisfied. In a level-clocked circuit, however, critical paths are more difficult to identify since they may vary from one clock period value to another. This was illustrated in [LE94] through the following example. Figure 10.14 shows a situation where three paths, p_1, p_2, and p_3, connect the vertex u to the vertex v; the delay of each node is written within its corresponding vertex. The description of each path is as follows:

Path	Vertices	$w(p)$	Path delay
p_1	$u \Rightarrow v_2 \Rightarrow v_3 \Rightarrow v_4 \Rightarrow v$	4	10
p_2	$u \Rightarrow v_1 \Rightarrow v$	2	3
p_3	$u \Rightarrow v_5 \Rightarrow v$	2	6

where $w(p)$ denotes the number of registers that currently lie on the path.

If a path from u to v has w registers on it, then for a symmetric two-phase clock with an active period of T_ϕ, the amount of time available for data to travel from u to v is $(w + 2) \cdot T_\phi$ time units, corresponding to the total active time available from the input register of node u to the output register of node v. This leads to the following statements:

- When a symmetric two-phase clock with a period of 3 units (and therefore, an active time of 1.5 units) is applied, we require that $w_r(p_1) = 5$ and $w_r(p_3) = 2$. Therefore, the path p_1 is critical since satisfying the constraint on p_3 in the given circuit would violate the constraint on p_1.

- When the period of the same clock is changed to 5 units, the requirements are altered to $w_r(p_1) = 2$ and $w_r(p_3) = 1$; in this case, satisfying the former would lead to a violation of the latter.

- It is easily verified that the crossover point for the critical paths occurs at a clock period of 4 units, where both are critical.

Note that the path p_2 is always subservient to path p_3 since it has the same number of registers, but a smaller delay, and can, hence, never be critical.

10.6.2 Retiming of single phase level-clocked circuits

In [SBS91] an MILP formulation is presented for retiming single phase level-clocked circuits. Constraints for correct clocking of single-phase circuits are laid down, which were essentially similar to the SMO constraints in Section 7.3.2. A new functionality constraint was also introduced in this work, which maintains temporal equivalence between the initial and the retimed circuit. The constraint identifies a set of fundamental cycles in the directed graph corresponding to this problem, and ensures that the sum of latches on each of these is maintained during retiming. The property of the set of fundamental cycles is that all other cycles can be expressed as a composition of these fundamental cycles. The essential idea of the approach is to build a spanning tree on the directed graph. If the graph has m edges and n vertices, then there are $m - n + 1$

edges that are excluded from the spanning tree. It is proved in [SSBS92] that each of these edges corresponds to a linearly independent cycle by showing that the cycle-edge incidence matrix has rank $m - n + 1$.

10.6.3 The TIM approach

TIM [Pap93, PR93b] presents polynomial time algorithms for retiming and clock tuning of two-phase level clocked circuits. Since the notation used in the work differs slightly from the SMO model, we will take a detour to define a few terms. A clocking scheme is defined as $\pi = \langle \phi_0, \gamma_0, \phi_1, \gamma_1 \rangle$, where P is the clock period, ϕ_i the duty cycle of phase i and γ_i is the time gap between the instant where the phase i clock goes low and the phase $i(i+1) \bmod k$ clock goes high. Each vertex, v, corresponds to a gate, and is associated with a phase, $\chi(v) \in \{0, 1\}$, i.e., $\phi_{\chi(v)}$ clocks the last latch on any edge that ends at v.

The constraint for proper clocking on a path $p : u \rightsquigarrow v$ is given by the following relations:

$$d(p) \leq P\left(\frac{1 + w(p)}{2}\right) + \phi_{\chi(u)}$$

$$+ P\lfloor \frac{r(v)}{2} \rfloor + (r(v) \bmod 2)\left(\gamma_{\chi(u)} + \phi_{1-\chi(u)}\right)$$

$$- P\lfloor \frac{r(u)}{2} \rfloor - (r(u) \bmod 2)\left(\gamma_{\chi(u)} + \phi_{\chi(u)}\right), \text{ if } \chi(u) \neq \chi(10.25)$$

$$d(p) \leq P\left(\frac{2 + w(p)}{2}\right) - \gamma_{1-\chi(u)}$$

$$+ P\lfloor \frac{r(v)}{2} \rfloor + (r(v) \bmod 2)\left(\gamma_{1-\chi(u)} + \phi_{\chi(u)}\right)$$

$$- P\lfloor \frac{r(u)}{2} \rfloor - (r(u) \bmod 2)\left(\gamma_{\chi(u)} + \phi_{\chi(u)}\right), \text{ if } \chi(u) = \chi(v)(10.26)$$

When $\chi(u) \neq \chi(v)$, $1 - \chi(u) = \chi(v)$ and the relation (10.25) enforces the constraint that the delay on any path must be less than the summation of the following terms (the interpretation of relation (10.26) is analogous):

- The first line on the right hand side is the timing allowance on the path p before retiming.

- The second line is the change in the timing allowance of the path p due to the latches that where moved either on or off p through gate v during the retiming process. The first term ($P\lfloor \frac{r(v)}{2} \rfloor$) is the number of complete clock periods added to the timing allowance, and the second term accounts for fractional periods added. If $r(v)$ is even, no fractional periods are added and the second term is zero ($r(v) \bmod 2 = 0$). If $r(v)$ is positive, latches are shifted on to path p and the timing allowance is increased; if it is negative, latches are being shifted off from path p and the timing allowance is decreased.

If $r(v)$ is odd, the fractional change must also be considered. For a positive value of $r(v)$, the last latch moved on p is of phase $1 - \chi(v)$. Therefore, the latch on the boundary of p (which includes gate v) will be of phase $1 - (1 - \chi(v)) = \chi(v)$ which is same as $1 - \chi(u)$. Thus, the fractional period added by this last latch is $\gamma_{\chi(u)} + \phi_{1-\chi(u)}$. If $r(v)$ is negative, then the last latch moved off p by forward retiming across v is of phase $\chi(v)$, and the quantity $\gamma_{1-\chi(v)} + \phi_{\chi(v)} = \gamma_{\chi(u)} + \phi_{1-\chi(u)}$ must be subtracted from the timing allowance of path p.

- The third line is the change in the timing allowance of the path p due to latches being moved off or on p through gate u, and is derived in a similar manner to the second line.

For a more formal mathematical proof, the reader is referred to [Pap93].

These constraints in relations (10.25) and (10.26), together with the nonnegativity constraint, $r(v) + w(e_{uv}) \geq r(u)$, form an integer monotonic program that can be solved using graph-based algorithms.

A similar method for retiming level-clocked circuits was independently presented in [LE91, LE92, LE94], and was based on the SMO formulation of timing constraints.

10.6.4   The relation between retiming and skew for level-clocked circuits

To derive timing constraints in the presence of skews, the SMO model of Section 7.3.2 is augmented with new notation. A skew, x_i, is associated with every latch $i \in \Psi$, where Ψ is the set of all latches in the circuit. It is worth pointing out at this juncture that the skew values here are not physical skews to be applied to the final circuit, but conceptual ideas that will eventually lead to achieve a retiming solution. No restrictions are placed on the value of x_i, i.e. $-\infty \leq x_i \leq \infty$.

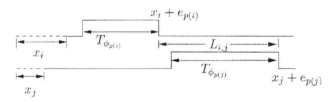

Figure 10.15.   The latch shift operator [MS99].

We define a *latch shift* operator $L_{i,j}$, shown in Figure 10.15, much like the phase shift operator in the SMO formulation. This operator converts time from the local time zone of latch i to the local time zone of latch j, taking into account their skews. It is defined as

$$L_{i,j} = \begin{cases} (x_j + e_{p(j)}) - (x_i + e_{p(i)}) & \text{for } i < j \\ T_\Phi + (x_j + e_{p(j)}) - (x_i + e_{p(i)}) & \text{for } i \geq j \end{cases}$$

which can be rewritten in terms of the phase shift operator as

$$L_{i,j} = (x_j - x_i) + E_{p(i),p(j)} \qquad (10.27)$$

In presence of skews at latches, the timing constraints in relation (7.17) and (7.20), must be modified by using the latch shift operator instead of the phase shift operator; only long path constraints are considered here. Thus, the timing constraints for a level clocked circuit to be properly clocked by a clock schedule, Φ, in presence of skews are

$$D_i + d(i,j) - L_{i,j} \leq D_j \quad \forall\, i \rightsquigarrow j \mid i,j \in \Psi$$
$$T_\Phi - T_{p(i)} \leq D_i \leq T_\Phi \quad \forall\, i \in \Psi$$

For simplicity, this discussion ignores the clock-to-Q delays and setup times, though these may easily be inserted into these equations. These timing constraints can be rewritten as

$$(x_i + D_i) + d(i,j) - E_{p(i),p(j)} \leq (x_j + D_j) \quad \forall\, i \rightsquigarrow j \mid i,j \in \Psi$$
$$T_\Phi - T_{p(i)} \leq D_i \leq T_\Phi \quad \forall\, i \in \Psi$$
$$-\infty \leq x_i \leq \infty \quad \forall\, i \in \Psi$$

To make the discussion simpler, we subtract T_Φ from both sides of the first relation, and substitute

$$X_i = (x_i + D_i - T_\Phi) \qquad (10.28)$$

The quantity X_i is referred to as the *Global Departure Time* (GDT). Therefore, we have

$$X_i + d(i,j) - E_{p(i),P(j)} \leq X_j \quad \forall\, i \rightsquigarrow j \mid i,j \in \Psi$$
$$-\infty \leq X_i \leq \infty \quad \forall\, i \in \Psi$$

These can be rewritten as the following set of difference constraints:

$$X_i - X_j \leq E_{p(i),p(j)} - d(i,j) \quad \forall\, i \rightsquigarrow j \mid i,j \in \Psi \qquad (10.29)$$
$$-\infty \leq X_i \leq \infty \quad \forall\, i \in \Psi$$

The difference constraint between GDT values of two latches given in relation (10.29) is similar to the difference constraints between skews at FF's in relation (10.8). This suggests a relation between retiming and GDT values of level-sensitive latches, similar to the retiming-skew relationship for edge triggered FF's. The following theorem is similar to the corresponding result for edge-triggered circuits:

(a) Retiming transformations may be used to move latches from all of the n inputs of any combinational block to all of its outputs. The equivalent GDT of the relocated latch at output j, considering long path constraints only, is given by

$$X_j = \max_{1 \leq i \leq n} (X_i + \overline{d}(i,j))$$

where the $X_i, 1 \leq i \leq n$, are the GDT's at the input latches, X_j is the equivalent GDT at output j, and $\bar{d}(i,j)$ is the worst-case delay of any path from i to j.

(b) Similarly, latches may be moved from all of the m outputs of any combinational block to all of its inputs, and the equivalent GDT at input k, considering long path constraints only, is given by

$$X_k = \min_{1 \leq j \leq m} (X_j - \bar{d}(k,j)) \qquad (10.30)$$

where the $X_j, 1 \leq j \leq m$, are the GDT's at the input latches, X_k is the equivalent GDT at input k, and $\bar{d}(k,j)$ is the worst-case delay of any path from k to j.

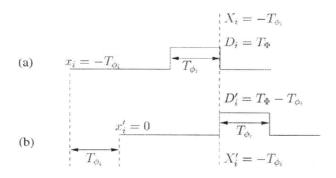

Figure 10.16.   The ability of a latch to absorb some skew [MS99].

The physical meaning of the GDT is as follows. If arbitrary skews could be applied at latches, the skew, x_i, of a latch could be adjusted so as to force $D_i = T_\Phi$, which is same as a negative edge triggered FF. Since $X_i = x_i + D_i - T_\Phi$, setting $D_i = T_\Phi$ gives $x_i = X_i$. Hence, we can conceive of the GDT, X_i, for a latch to be similar to the skew for an FF.

Note that in reality, we are not compelled to set $D_i = T_\Phi$, and that D_i can be reduced by as much as T_{ϕ_i}, if x_i is increased by the same amount while keeping X_i constant. Since only GDT's, X_i, appear in the timing constraint of relation (10.29), keeping them constant keeps the clock period constant. As an illustration of this idea, consider Figure 10.16(a) where $x_i = -T_{\phi_i}$ and $D_i = T_\Phi$; this implies that $X_i = -T_{\phi_i}$. The skew here may be increased to zero $(x_i' = 0)$, without changing the GDT, as shown in Figure 10.16(b), by reducing the departure time by the same amount $D_i' = T_\Phi - T_{\phi_i}$, thereby leaving the GDT unchanged $(X_i' = X_i = -T_{\phi_i})$. Therefore, a skew of up to $-T_{\phi_i}$ can be absorbed in the D_i without violating the long path constraint. Thus, any GDT value between $-T_{\phi_i}$ and 0 is permissible, and this range will be referred to as the allowable range. If different phases have different active intervals, then the allowable GDT range of a latch will depend on its phase. Therefore, in this model, level-sensitive latches can be conceptualized as FF's that have the capacity to absorb some skew.

At this time, we also note the relation between the GDT, X_i, of a latch i and the corresponding minimum magnitude skew, x_i:

$$
x_i = \begin{cases} X_i & \text{if } X_i \geq 0 \\ 0 & \text{if } -T_{\phi_i} \leq X_i \leq 0 \\ X_i + T_{\phi_i} & \text{if } -T_{\phi_i} > X_i \end{cases} \tag{10.31}
$$

10.6.5 Minimum period retiming

The relationship between GDT and retiming presented in Section 10.6.4 can be utilized to map the problem of retiming level-clocked circuits for minimum period to the simpler problem of retiming edge-triggered circuits for minimum period as solved in Section 10.4.2. This mapping motivates the following two-phase method of retiming for minimum clock period under a specified clocking scheme. The execution times for such an implementation are fast and a 50,000 gate circuit was shown to be retimed in about seven minutes in a computer of mid-to-late 90s vintage.

The two phases in this method mirror those for edge-triggered circuits, and are as follows:

Phase A: In phase A, the clock period, T_Φ, is minimized, and a set of GDT values that will achieve this period is determined. For a specified clocking scheme can be represented as the following linear program:

$$
\begin{aligned}
&\text{minimize} && T_\Phi \\
&\text{subject to} && X_i - X_j \leq E_{p(i),p(j)} - d(i,j) \quad i \rightsquigarrow j \mid i,j \in \Psi \quad (10.32)
\end{aligned}
$$

As before, the LP is solved through a binary search on the clock period, checking for feasibility at each point in the search as was done in Section 9.3.3.

Phase B: In Phase B, the GDT values obtained in Phase A are reduced by applying transformations that alter the GDT through retiming, using techniques similar to those in Section 10.4.2. This procedure relocates the latches with nonzero GDT's across logic gates, while maintaining the optimal clock period previously found. Because of the freedom provided to D_i by the active interval of clock phase $p(i)$, which allows D_i to be set to any value between $T_\Phi - T_{\phi_{p(i)}}$ and T_Φ, $x_i = 0$ can be achieved if $-T_{\phi_{p(i)}} \leq X_i \leq 0$. If x_i cannot be set to zero, then X_i is brought as close to 0 or $-T_{\phi_{p(i)}}$ as possible so as to minimize the magnitude of the final skew x_i.

At the end of the retiming procedure in Phase B, the magnitude of skew at each latch i, is no more than

$$
\gamma_{p(i)} = \max\left(0, \frac{M - T_{\phi_{p(i)}}}{2}\right)
$$

where M is the maximum delay of any gate in the circuit. Thus, if the maximum gate delay is less than the least T_{ϕ_i}, the the optimal skew optimization period can always be achieved by retiming. At this time, if all skews in a k-phase circuit are set to zero, then the final clock period (P_r) satisfies the following condition

$$P_r \leq P_s + \sum_{i=1}^{k} \max(0, M - T_{\phi_i})$$

where P_s is the optimal clock period with skews found in Phase A, and M is the maximum delay of any gate in the circuit.

The advantage of this method is that it constructs a small and sparse graph only once.

We first describe the two phases of minimum period retiming, followed by the special case of retiming a circuit for a specified clock period. We then present the bound on the difference between the optimal skew optimization period and the clock period obtained by our method.

10.7   MINIMUM AREA RETIMING OF LEVEL-CLOCKED CIRCUITS

The minimum area retiming LP for level-clocked circuits is similar to the LP for edge-triggered circuits given in (10.17). Unfortunately, under general clock schedules with unequal phases, it must be modeled as a general integer linear program of the type given in [LE91], while restricting the clock scheme to a symmetric multi-phase clock enables us to model the minimum area retiming problem as an efficiently solvable LP (dual of min-cost flow problem) [PR93b].

Like the minimum area LP of (10.17), the minimum area LP for level-clocked circuits also contains circuit constraints, period constraints and mirror constraints. The circuit and mirror constraints are defined in the same way as in Section 10.5.1, although the period constraints are somewhat different since the timing constraints in level-clocked circuits differ from those in edge-triggered circuit. Moreover, constraint pruning becomes more complex and requires fresh strategies. A detailed discussion of this approach, called the Minaret-L algorithm, is presented in [MS99].

10.8   SUMMARY

In this chapter, a detailed treatment of retiming algorithms has been provided. Retiming is a technique for purely sequential timing optimization, which allows cycle borrowing by moving the flip-flops or latches at cycle boundaries. Algorithms for retiming both edge-triggered and level-clocked circuits have been reviewed.

Most of the issues discussed here are related to retiming combinational logic. In recent years, wire delays have increased to the level where it may take several clock cycles even for a highly optimized wire to go from one end of a chip to the other, and this problem is only projected to become worse in the future [SMCK03]. As a consequence, it is becoming essential to pipeline wires in a circuit to ensure correct timing behavior. Several new approaches to this problem have been presented in the recent past [HAT02, Coc02], and although it is too early to say so definitively, it appears that the use of retiming-like concepts [LZ03, NS04] may hold great promise for solving problems in this area.

Notes

1. The calculation of indices i here span the range $0, \cdots, k$, and in this discussion, $i - 1$ is calculated modulo k.

2. This result is easy to see on this simple example. The reader is referred to Chapter 9 for a general procedure for finding the optimal skews for a general circuit

11 CONCLUSION

This book has attempted to provide an overview of techniques that are used in the timing analysis of digital circuits, with an exposition of methods used for analyzing circuits at the gate/interconnect level, then at the level of a combinational stage, and finally as a larger sequential circuit. Methods for timing optimization have been discussed, through transistor sizing and V_t optimization at the transistor level for combinational circuits, and through clock skew scheduling and retiming for sequential circuits. A survey of statistical timing methods has also been presented.

While this book covers several of the basics of timing analysis, there are several topics that it does not cover explicitly in great detail. However, the foundation for understanding these issues is well laid out in our discussion.

Inductive effects are becoming more prominent in nanometer technologies, particularly for long global wires such as those used for clock distribution, global busses, and power supply nets. There is a great deal of literature available on the problem of extracting and analyzing inductive networks, and the material in Chapters 2 and 3 is very useful in understanding these issues.

Timing-driven synthesis Logic synthesis has traditionally been divided into the technology-independent and technology-dependent phases. The latter phase, which may perform technology mapping to a library, explicitly uses timing information for synthesis. However, such timing estimates are closely related to the estimated parasitics. Previously, wire load models were considered to be sufficient for this purpose, but increasingly, there is a need for more accurate models. This step must perform timing analysis in the absence of complete design information, and may make it up using a physical prototype, or perhaps a probabilistic model for the distribution of net parasitics.

Layout-driven timing optimization Timing optimization during physical design can be performed during the stages of placement, and routing. Placement algorithms may use either simple estimates for timing that are net-based, attempting to limit the size or delay of each net, or be more intelligent and use path-based metrics, which attempt to limit the delay on the path. The advantage of the latter is that they can permit nets that are not on critical paths to be longer, and can dynamically evaluate which paths are more critical; however, they are also more computational. However, there is a growing realization that some sort of explicit timing optimization must be interleaved with placement, and timing analysis methods play a vital role in deciding how this is to be performed. The fast timing metrics discussed in Chapter 3 are likely to play an important role in these developments.

Timing issues also play an important role in routing. Global wires must be routed in such a way that they obey constraints on timing and congestion using accurate delay metrics. In addition, the effect of noise on delay, and coupling effects due to crosstalk must be incorporated both in global and detailed routing. The methods of Section 4.5 can provide a basis for such optimizations.

Timing closure The timing closure problem requires timing estimates to be taken into account in all stages of design, and can be thought of as a conglomeration of methods for timing-driven synthesis (behavioral synthesis or logic synthesis), timing-driven placement and timing-driven routing, into a flow that guarantees timing closure with as little iteration as possible. Clearly, methods for the measurement and optimization of timing play an important role in implementing such a flow.

Since the field of timing analysis and optimization is an active and fertile research area, it is likely that new results will continue to be published in the future, and the reader is referred to conferences such as the ACM/IEEE Design Automation Conference (DAC) and the IEEE/ACM International Conference on Computer-Aided Design (ICCAD), and to journals such as the IEEE Transactions on Computer-Aided Design of Integrated Circuits and Systems, the IEEE Transactions on VLSI Systems, and the ACM Transactions on Design Automation of Electronic Systems.

Appendix A
Closed-form formula for the roots of cubic and quartic equations

This appendix reproduces results on the solution of cubic and quartic equations [Ric66].

A.1 CUBIC EQUATIONS

A general cubic equation is given by

$$x^3 + bx^2 + cx + d = 0$$

Setting $x = y - b/3$, this becomes

$$y^3 + py + q = 0$$

where $p = c - \frac{b^2}{3}$ and $q = d - \frac{bc}{3} + \frac{2b^3}{27}$. Further substituting $y = z - \frac{p}{3z}$, this reduces to

$$z^3 - \frac{p^3}{27z^3} + q = 0$$

$$\text{i.e., } z^6 + qz^3 - \frac{p^3}{27} = 0$$

This is a quadratic in z^3, which may be solved to yield six roots for z, namely, α, $\omega\alpha$, $\omega^2\alpha$, β, $\omega\beta$, and $\omega^2\beta$, where

$$\alpha = sqrt[3]\frac{q}{2} + \sqrt{\frac{q^2}{4} + \frac{p^3}{27}}$$

$$\beta = sqrt[3]\frac{q}{2} - \sqrt{\frac{q^2}{4} + \frac{p^3}{27}}$$

$$\omega = -\frac{1}{2} + \frac{\sqrt{3}}{2}i \text{ (the cube roots of 1)}$$

A.2   QUARTIC EQUATIONS

A general quartic equation is given by

$$x^4 + bx^3 + cx^2 + dx + e = 0$$

Substituting $x = y - \frac{b}{4}$, we get an equation of the form

$$y^4 + gy^2 + hy + k \;=\; 0$$

$$\text{where } g \;=\; c - \frac{3b^2}{8}$$

$$h \;=\; \frac{b^3}{8} - \frac{bc}{2} + d$$

$$k \;=\; \frac{b^2 c}{16} - \frac{3b^4}{256} - \frac{bd}{4} + e$$

This may be rewritten as

$$\left(y^2 + \frac{z}{2}\right)^2 - \left[(z - g)y^2 - hy + \left(\frac{z^2}{4} - k\right)\right] = 0$$

where z is an as yet undetermined constant. We will choose z to make the second term a perfect square, substituting $p^2 = \left[(z - g)y^2 - hy + \left(\frac{z^2}{4} - k\right)\right]$. For this relation to be true, the discriminant of the quadratic must be zero. In other words,

$$h^2 - 4(z - g)\left(\frac{z^2}{4} - k\right) \;=\; 0$$

$$\text{i.e., } z^3 - gz^2 - 4kz - (h^2 - 4gk) \;=\; 0 \tag{A.1}$$

Equation A.1 is a cubic that can be solved as explained in Section A.1. Given any root of this, say z^* (this is where z is determined!), the original equation becomes

$$\left(y^2 + \frac{z^*}{2}\right)^2 - \left(y\sqrt{z_1 - g} - \frac{h}{2\sqrt{z_1 - g}}\right)^2 \;=\; 0$$

$$\text{i.e., } y^2 + \frac{z^*}{2} + y\sqrt{z_1 - g} - \frac{h}{2\sqrt{z_1 - g}} \;=\; 0$$

$$\text{and } y^2 + \frac{z^*}{2} - y\sqrt{z_1 - g} + \frac{h}{2\sqrt{z_1 - g}} \;=\; 0$$

These two quadratics may be solved to find the four roots of the quartic equation. Note that due to self-consistency, three of the roots will coincide with those already found by solving the intermediate cubic equation.

Appendix B
A Gaussian approximation of the PDF of the maximum of two Gaussians

Although the maximum of two Gaussians is not a Gaussian, it is possible to approximate it as such, and a method for doing so was suggested in [Cla61]. If ξ and η are two Gaussian random variables, $\xi \sim (\mu_1, \sigma_1)$, $\eta \sim (\mu_2, \sigma_2)$, with a correlation coefficient of $r(\xi, \eta) = \rho$, then the mean μ_t and the variance σ_t^2 of $t = \max(\xi, \eta)$ can be approximated by:

$$\mu_t = \mu_1 \cdot \Phi(\beta) + \mu_2 \cdot \Phi(-\beta) + \alpha \cdot \varphi(\beta) \tag{B.1}$$

$$\sigma_t^2 = (\mu_1^2 + \sigma_1^2) \cdot \Phi(\beta) + (\mu_2^2 + \sigma_2^2) \cdot \Phi(-\beta)$$
$$+ (\mu_1 + \mu_2) \cdot \alpha \cdot \varphi(\beta) - \mu_t^2 \tag{B.2}$$

where

$$\alpha = \sqrt{\sigma_1^2 + \sigma_2^2 - 2\sigma_1\sigma_2\rho} \tag{B.3}$$

$$\beta = \frac{(\mu_1 - \mu_2)}{\alpha} \tag{B.4}$$

$$\varphi(x) = \frac{1}{\sqrt{2\pi}} exp\left[-\frac{x^2}{2}\right] \tag{B.5}$$

$$\Phi(x) = \frac{1}{\sqrt{2\pi}} \int_{-\infty}^{x} exp\left[-\frac{y^2}{2}\right] dy \tag{B.6}$$

The formula will not apply if $\sigma_1 = \sigma_2$ and $\rho = 1$. However, in this case, the *max* function is simply identical to the random variable with largest mean value.

Moreover, from [Cla61], if γ is another normally distributed random variable and $r(\xi, \gamma) = \rho_1$, $r(\eta, \gamma) = \rho_2$, then the correlation between γ and $t = \max(\xi, \eta)$ can be obtained by:

$$r(t, \gamma) = \frac{\sigma_1 \cdot \rho_1 \cdot \Phi(\beta) + \sigma_2 \cdot \rho_2 \cdot \Phi(-\beta)}{\sigma_t} \tag{B.7}$$

Using the formula above, we can find all the values needed. As an example, let us see how this can be done by first starting with a two-variable *max* function, $d_{max} = max(d_i, d_j)$. Let d_{max} be of the form

$$d_{max} = \mu_{d_{max}} + a_1 p_1' + \cdots + a_m p_m' \tag{B.8}$$

We can find the approximation of d_{max} as follows:

1. Given the expressions that express each of d_i and d_j as linear combinations of the principal components, determine their mean and sigma values μ_{d_i}, σ_{d_i}, μ_{d_j}, and σ_{d_j}, respectively; recall that $\sigma_d^2 = \sum_{i=1}^{m} k_i^2$.

2. Find the correlation coefficient between d_i and d_j where $cov(d_i, d_j)$, the covariance of d_i and d_j can be computed, since $cov(d_i, d_j) = \sum_{r=1}^{m} k_{ir} k_{jr}$.

   Now if $r(d_i, d_j) = 0$ and $\sigma_{d_i} = \sigma_{d_j}$, set d_{max} to be identical to d_i or d_j, whichever has larger mean value and we can stop here; otherwise, we will continue to the next step.

3. Calculate the mean $\mu_{d_{max}}$ and variance $\sigma^2_{d_{max}}$ of d_{max} using Equations (B.1) and (B.1).

4. Find all coefficients a_r of p'_r. Note that $a_r = cov(d_{max}, p'_r)$, $cov(d_i, p'_r) = k_{ir}$, and $cov(d_j, p'_r) = k_{jr}$. Applying Equation (B.7), the values of $cov(d_{max}, p'_r)$ and thus a_r can be calculated.

5. After all of the a_r's have been calculated, determine $s_0 = \sqrt{\sum_{r=1}^{m} a_r^2}$. Normalize the coefficient by resetting each $a_r = a_r \frac{\sigma_{d_{max}}}{s_0}$.

Several useful extensions are possible:

- The minimum of two random variables ξ and η may be computed by using the above procedure to compute $-\max(-\xi, -\eta)$.

- The calculation of the two-variable max function can easily be extended for an n-variable max function by repeating the steps of the two-variable case recursively.

Appendix C
On the convexity of the generalized posynomial formulation for transistor sizing

C.1 PROOF OF CONVEXITY

The objective functions of area and power can be modeled as posynomials in the usual ways. This appendix discusses the proof that the generalized posynomial formulation is equivalent to a convex program under the timing and ratioing constraints.

The ensuing discussion first shows that the delays of individual paths satisfy the property of convexity, and uses this fact to prove the convexity of the optimization problem. It is to be emphasized that this discussion is purely for expository purposes; the optimizer used in this work for sizing *does not* require the enumeration of all paths, and performs the optimization efficiently by checking, through a timing analysis, whether the constraints are satisfied or not. For details, the reader is referred to [SRVK93].

Let the critical path of the circuit be represented by a set of stages, where each stage represents a gate. Let us first consider a scenario with fully characterized gates where no primitives are used, but the delay is characterized in terms of the size of each transistor. Then, substituting the characterization variables explicitly into Equation (8.29), we see that the fall delay of the gate corresponding to stage i has the following form:

$$\text{Delay}_i \;=\; \sum_i P_i \cdot (w_{n1}^{-1} + c_{n1})^{\beta_1} \cdots (w_{nm_n}^{-1} + c_{nm_n})^{\beta_{m_n}} \cdot$$
$$(w_{p1} + c_{p1})^{\beta_1} \cdots (w_{pm_p} + c_{pm_p})^{\beta_{m_p}} \cdot (\tau_{i-1} + c_\tau)^{\beta_\tau} \Pi_j (C_j + c_{C_j})^{\beta_{C_j}}$$

$$(C.1)$$

and the output fall transition time of the gate in stage i has the form.

$$\tau_i \;=\; Q \cdot (w_{n1}^{-1} + k_{n1})^{\gamma_{n1}} \cdots (w_{nm_n}^{-1} + k_{nm_n})^{\gamma_{nm_n}} \cdot$$
$$(w_{p1} + k_{p1})^{\gamma_{p1}} \cdots (w_{pm_p} + k_{pm_p})^{\gamma_{pm_p}} (\tau_{i-1} + k_\tau)^{\gamma_\tau} \Pi_j (C_j + k_{C_j})^{\gamma_{C_j}}$$

$$(C.2)$$

where $P_i > 0$, $Q > 0$, c_{ni}, c_{pi}, k_{ni}, k_{pi}, β_{ni}, β_{pi}, γ_{ni}, γ_{pi} $\forall\, i$, k_{C_j}, c_{C_j} $\forall\, j$, k_τ, c_τ, β_{C_j}, γ_{C_j}, β_τ, γ_τ are real constants. The w_{ni} and w_{pi} values, as usual, refer to the nmos and pmos transistor sizes, τ refers to the transition time, and the C_j's correspond to the capacitances at the gate output and at internal nodes. The rise delay and rise transition time expressions are similar, with the roles of w_n and w_p interchanged. We will show that the delay and transition time functions have the form of generalized posynomials.

The capacitance at each internal or gate output node i, C_i is modeled by

$$C_i = \sum_j k'_j w_j + k''$$

$$(C.3)$$

where the k_j' and k'' values are real constants, and w_j's represent the equivalent transistor widths in the circuit.

Equations (C.2) and (C.3) can be recursively substituted in Equation (C.1), back to the primary inputs (where the transition time is a fixed constant) to express the delay entirely as a function of the w variables, resulting in an expression of the form

$$\text{Delay} \quad = \quad \sum K(\sum_{j=1}^{m} a_{1j} w_j^{\Delta} + b_1)^{\beta_1'} (\sum_{j=1}^{m} a_{2j} w_j^{\Delta} + b_2)^{\beta_2'} (\sum_{j=1}^{m} a_{nj} w_j^{\Delta} + b_n)^{\beta_n'}$$

(C.4)

where a_{ij}'s, b_i's, c_{ij}'s, d_i's are real constants, Δ = either -1 or 1, $K > 0$, and the $\beta_i' > 0 \; \forall i$. Note that the transition time τ at the primary input is absorbed in K.

It is easily verified that the path delay is a generalized posynomial function of the device widths that can be mapped on to a convex function. The objective function is chosen as a weighted sum of the transistor sizes, which is clearly a generalized posynomial form. Using identical arguments to [FD85, SRVK93], since the maximum of convex functions is convex, the problem of area minimization under delay constraints for "template" gates can be shown to be a convex programming problem.

The problem of power minimization can be dealt with in a similar fashion; since the edge rates are being explicitly controlled in this formulation, the short-circuit power is implicitly controlled and, unlike [SC95, BOI96] can be neglected from the cost function. As a result, the power objective merely requires minimization of the dynamic power, which is well known to be a weighted sum of the device sizes [SC95, BOI96].

For gates that do not adhere to the template, the mapping techniques described in Section 8.5 may be used to model the delay function. We will now show that in such a case, the delay function continues to remain in the generalized posynomial form. Let $w_1', \cdots, w_m'$ represent transistor widths in the primitives the gates are mapped to. In the process of mapping the gates, the transistor widths in the primitives can be expressed in terms of the actual transistor widths in the circuit. Let $w_1, \cdots, w_n$ represent the actual transistor widths in the circuit. Then w''s can be expressed as

$$w_i'^{-1} = \sum_{q \in \{1 \cdots n\}} w_q^{-1}, 1 \leq i \leq m$$

(C.5)

All occurrences of value of $w_i'^{-1}$, which is a basic variable in the characterization equation can be substituted as above in Equation (C.4), maintaining the generalized posynomial property of the delay equation.

With regard to the ratioing constraints, each left hand side in the inequalities listed in relation (8.31) is a generalized posynomial since (using a similar argument as above), each denominator term of $w_i'^{-1}$ is substituted by a sum of w_i^{-1} values, and the numerator terms correspond to sums of w_i's. Therefore, these constraints maintain the convexity properties of the problem formulation.

C.2   RELATION OF A GENERALIZED POSYNOMIAL PROGRAM TO A POSYNOMIAL PROGRAM

Traditionally posynomial programs have been solved by using geometric optimizers. In this work, we have used generalized posynomials to model the gate delay and power dissipation. While these functions are provably equivalent to convex functions, a precise relationship between generalized posynomial programming problems and posynomial programming problems would permit the use of geometric optimizers to solve the problem formulated here. In this section, we describe a technique for transforming the set of delay constraints described by generalized posynomials into a posynomial form.

As will be shown shortly, the transformation is carried out by the introduction of additional variables. Consider the constrained order 1 generalized posynomial constraint below:

$$\sum_i \gamma_i \prod_{j=1}^m \left(\sum_{l=1}^{p_i} \omega_{ijl} \prod_{s=1}^n x_s^{a_{ijls}} \right)^{\beta_{ij}} \leq D_{req} \tag{C.6}$$

where D_{req} is a constant and $\beta_{ij} \geq 0$. The term in parentheses is a posynomial, and if we set it to the variable y_j, the above relation may be rewritten as

$$\sum_i \gamma_i \prod_{j=1}^m y_j^{\beta_{ij}} \leq D_{req} \tag{C.7}$$

$$\sum_{l=1}^{p_i} \omega_{ijl} \prod_{s=1}^n x_s^{a_{ijls}} = y_j \tag{C.8}$$

The latter equality can, of course, be written as a pair of inequalities. We now make use of a subtle observation: if we relax this equality to just one "$\leq$" inequality, then for any variable assignment that satisfies the relaxed set of constraints, since $\gamma_i > 0$ and $\beta_{ij} \geq 0$, it must be true that

$$\sum_i \gamma_i \prod_{j=1}^m \left(\sum_{l=1}^{p_i} \omega_{ijl} \prod_{s=1}^n x_s^{a_{ijls}} \right)^{\beta_{ij}} \leq \sum_i \gamma_i \prod_{j=1}^m y_j^{\beta_{ij}} \tag{C.9}$$

In conjunction with constraint (C.7), this implies that constraint (C.6) must be satisfied.

Therefore, a generalized posynomial constraint of order 1 can be replaced by the posynomial constraints

$$\sum_i \gamma_i \prod_{j=1}^m y_j^{\beta_{ij}} \leq D_{req} \tag{C.10}$$

$$\sum_{l=1}^{p_i} \omega_{ijl} \prod_{s=1}^n x_s^{a_{ijls}} \leq y_j \tag{C.11}$$

This substitution method can be used, in general, for generalized posynomials of order k. Such a procedure can reduce this to an order $k - 1$ generalized posynomial, and the substitution can be invoked recursively until a set of purely posynomial constraints is obtained.

References

[AB98]   H. Arts and M. Berkelaar. Combining logic synthesis and retiming.
            In *Workshop Notes, International Workshop on Logic Synthesis*,
            pages 136–139, 1998.

[ABP01]   R. Arunachalam, R. D. Blanton, and L. T. Pileggi. False cou-
            pling interactions in static timing analysis. In *Proceedings of the
            ACM/IEEE Design Automation Conference*, pages 726–731, 2001.

[ABZ⁺02]   A. Agarwal, D. Blaauw, V. Zolotov, S. Sundareswaran, M. Zhao,
            K. Gala, and R Panda. Path-based statistical timing analy-
            sis considering inter- and intra-die correlations. In *Proceedings
            of ACM/IEEE International Workshop on Timing Issues in the
            Specification and Synthesis of Digital Systems (TAU)*, pages 16–
            21, December 2002.

[ABZ03a]   A. Agarwal, D. Blaauw, and V. Zolotov. Statistical clock skew
            analysis considering intra-die process variations. In *Proceedings
            of the IEEE/ACM International Conference on Computer-Aided
            Design*, pages 914–921, 2003.

[ABZ03b]   A. Agarwal, D. Blaauw, and V. Zolotov. Statistical timing analysis
            for intra-die process variations with spatial corelations. In *Proceed-
            ings of the IEEE/ACM International Conference on Computer-
            Aided Design*, pages 900–906, November 2003.

[ABZ⁺03c]   A. Agarwal, D. Blaauw, V. Zolotov, S. Sundareswaran, M. Zhao,
            K. Gala, and R. Panda. Statistical timing analysis considering
            spatial correlations. In *Proceedings of the Asia/South Pacific De-
            sign Automation Conference*, pages 271–276, January 2003.

[ABZV03a]   A. Agarwal, D. Blaauw, V. Zolotov, and S. Vrudhula. Compu-
            tation and refinement of statistical bounds on circuit delay. In
            Proceedings of the ACM/IEEE Design Automation Conference,
            pages 348–353, June 2003.

[ABZV03b] A. Agarwal, D. Blaauw, V. Zolotov, and S. Vrudhula. Statistical timing analysis using bounds. In *Proceedings of Design, Automation and Test in Europe Conference*, pages 62–67, February 2003.

[ACI03] C. Amin, M. H. Chowdhury, and Y. I. Ismail. Realizable RLCK circuit crunching. In *Proceedings of the ACM/IEEE Design Automation Conference*, pages 226–231, 2003.

[ADK01] C. J. Alpert, A. Devgan, and C. V. Kashyap. RC delay metrics for perforamnce optimization. *IEEE Transactions on Computer-Aided Design of Integrated Circuits and Systems*, 20(5):571–582, May 2001.

[ADP02] E. Acar, F. Dartu, and L. T. Pileggi. TETA: Transistor-level waveform evaluation for timing analysis. *IEEE Transactions on Computer-Aided Design of Integrated Circuits and Systems*, 21(5):605–616, May 2002.

[ADQ98] C. J. Alpert, A. Devgan, and S. T. Quay. Buffer insertion for noise and delay optimization. In *Proceedings of the ACM/IEEE Design Automation Conference*, pages 362–367, 1998.

[ADQ99] C. J. Alpert, A. Devgan, and S. T. Quay. Buffer insertion with accurate gate and interconnect delay computation. In *Proceedings of the ACM/IEEE Design Automation Conference*, pages 479–484, 1999.

[AMH91] H. L. Abdel-Malek and A.-K. S. O. Hassan. The ellipsoidal technique for design centering and region approximation. *IEEE Transactions on Computer-Aided Design of Integrated Circuits and Systems*, 10(8):1006–1014, August 1991.

[AMO93] R. K. Ahuja, T. L. Magnanti, and J. B. Orlin. *Network Flows Theory, Algorithms and Applications*. Prentice Hall, Englewood Cliffs, NJ, 1993.

[ARP00] R. Arunachalam, K. Rajagopal, and L. T. Pileggi. TACO: Timing Analysis with COupling. In *Proceedings of the ACM/IEEE Design Automation Conference*, pages 266–269, 2000.

[ASB03] K. Agarwal, D. Sylvester, and D. Blaauw. An effective capacitance based driver output model for on-chip RLC interconnects. In *Proceedings of the ACM/IEEE Design Automation Conference*, pages 376–381, 2003.

[AZB03] A. Agarwal, V. Zolotov, and D. T. Blaauw. Statistical timing analysis using bounds and selective enumeration. *IEEE Transactions on Computer-Aided Design of Integrated Circuits and Systems*, 22(9):1243–1260, September 2003.

[BB98]   D. W. Bailey and B. J. Benschneider. Clocking design and analysis for a 600-MHz Alpha microprocessor. *IEEE Journal of Solid-State Circuits*, 33(11):1627–1633, November 1998.

[BBA+03]   M. R. Becer, D. Blaauw, I. Algolu, R. Panda, C. Oh, and I. N. Hajj V. Zolotov. Post-route gate sizing for crosstalk noise reduction. In *Proceedings of the ACM/IEEE Design Automation Conference*, pages 954–957, 2003.

[BC96]   A. Balakrishnan and S. Chakradhar. Retiming with logic duplication transformation: Theory and an application to partial scan. In *Proceedings of the International Conference on VLSI Design*, pages 296–302, 1996.

[Bel95]   K. P. Belkhale. Timing analysis with known false sub-graphs. In *Proceedings of the IEEE/ACM International Conference on Computer-Aided Design*, pages 736–740, 1995.

[Ber97]   M. Berkelaar. Statistical delay calculation, a linear time method. In *Proceedings of ACM/IEEE International Workshop on Timing Issues in the Specification and Synthesis of Digital Systems (TAU)*, pages 15–24, December 1997.

[BJ90]   M. R. C. M. Berkelaar and J. A. G. Jess. Gate sizing in MOS digital circuits with linear programming. In *Proceedings of the European Design Automation Conference*, pages 217–221, 1990.

[BJS77]   M. S. Bazaraa, J. J. Javis, and H.D. Sherali. *Linear Programming and Network Flows*. John Wiley, New York, NY, 1977.

[BK92]   K. D. Boese and A. B. Kahng. Zero-skew clock routing trees with minimum wirelength. In *Proceedings of the IEEE International ASIC Conference*, pages 17–21, 1992.

[BK01]   J. Baumgartner and A. Kuehlmann. Min-area retiming on flexible circuit structures. In *Proceedings of the IEEE/ACM International Conference on Computer-Aided Design*, pages 176–182, 2001.

[BKMR95]   K. Boese, A. Kahng, B. McCoy, and G. Robins. Near-optimal critical sink routing tree constructions. *IEEE Transactions on Computer-Aided Design of Integrated Circuits and Systems*, 14(12):1417–1436, December 1995.

[BOI96]   M. Borah, R. M. Owens, and M. J. Irwin. Transistor sizing for low power CMOS circuits. *IEEE Transactions on Computer-Aided Design of Integrated Circuits and Systems*, 15(6):665–671, June 1996.

[BP01]   M. W. Beattie and L. T. Pileggi. Inductance 101: Modeling and extraction. In *Proceedings of the ACM/IEEE Design Automation Conference*, pages 323–328, 2001.

[BPD00]   D. Blaauw, R. Panda, and A. Das. Removing user specified false paths from timing graphs. In *Proceedings of the ACM/IEEE Design Automation Conference*, pages 270–273, 2000.

[BSM94]   T. M. Burks, K. A. Sakallah, and T. N. Mudge. Optimization of critical paths in circuits with level-sensitive latches. In *Proceedings of the IEEE/ACM International Conference on Computer-Aided Design*, pages 468–473, 1994.

[BSM95]   T. M. Burks, K. A. Sakallah, and T. N. Mudge. Critical paths in circuits with level-sensitive latches. *IEEE Transactions on VLSI Systems*, 3(2):273–291, June 1995.

[BSO03]   D. Blaauw, S. Sirichotiyakul, and C. Oh. Driver modeling and alignment for worst-case delay noise. *IEEE Transactions on Computer-Aided Design of Integrated Circuits and Systems*, 22(4):157–166, April 2003.

[BVBD97]   L. Benini, P. Vuillod, A. Bogliolo, and G. De Micheli. Clock skew optimization for peak current reduction. *Journal of VLSI Signal Processing Systems for Signal, Image and Video Technology*, 16(2/3):117–130, June/July 1997.

[BZS02]   D. Blaauw, V. Zolotov, and S. Sundareswaran. Slope propagation in static timing analysis. *IEEE Transactions on Computer-Aided Design of Integrated Circuits and Systems*, 21(10):1180–1195, October 2002.

[CaPVS01]   J. Cong and D. Z. Pand an P. V. Srinivas. Improved crosstalk modeling for noise-constrained interconnect optimization. In *Proceedings of the Asia/South Pacific Design Automation Conference*, pages 373–378, 2001.

[CCH+98]   A. R. Conn, P. K. Coulman, R. A. Haaring, G. L. Morrill, C. Visweswariah, and C. W. Wu. JiffyTune: Circuit optimization using time-domain sensitivities. *IEEE Transactions on Computer-Aided Design of Integrated Circuits and Systems*, 17(12):1292–1309, December 1998.

[CCW99]   C.-P. Chen, C. C. N. Chu, and D. F. Wong. Fast and exact simultaneous gate and wire sizing by lagrangian relaxation. *IEEE Transactions on Computer-Aided Design of Integrated Circuits and Systems*, 18(7):1014–1025, July 1999.

[CEW+99]   A. R. Conn, I. M. Elfadel, W. W. Molzen, Jr., P. R. O'Brien, P. N. Strenski, C. Visweswariah, and C. B. Whan. Gradient-based optimization of custom circuits using a static-timing formulation. In *Proceedings of the ACM/IEEE Design Automation Conference*, pages 452–459, 1999.

[CGB01]   L.-C. Chen, S. K. Gupta, and M. A. Breuer. A new gate delay model for simultaneous switching and its applications. In *Proceedings of the ACM/IEEE Design Automation Conference*, pages 289–294, 2001.

[CGT92]   A. R. Conn, N. I. M. Gould, and P. L. Toint. *LANCELOT: A Fortran package for large-scale nonlinear optimization (Release A)*. Springer-Verlag, Heidelberg, Germany, 1992.

[CHA+92]   J.-H. Chern, J. Huang, L. Arledge, P.-C. Li, and P. Yang. Multilevel metal capacitance models for CAD design synthesis systems. *IEEE Electron Device Letters*, 13(1):32–34, January 1992.

[CHH92]   T. H. Chao, Y. C. Hsu, and J. M. Ho. Zero-skew clock net routing. In *Proceedings of the ACM/IEEE Design Automation Conference*, pages 518–523, 1992.

[CHKM96]   J. Cong, L. He, C. K. Koh, and P. H. Madden. Performance optimization of VLSI interconnect layout. *Integration: The VLSI Journal*, 21(1-2):1 94, November 1996.

[CHNR93]   E. Chiprout, H. Heeb, M. S. Nakhla, and A. E. Ruehli. Simulating 3-D retarded interconnect models using complex frequency hopping (CFH). In *Proceedings of the IEEE/ACM International Conference on Computer-Aided Design*, pages 66–72, 1993.

[Cir87]   M. A. Cirit. Transistor sizing in CMOS circuits. In *Proceedings of the 24th ACM/IEEE Design Automation Conference*, pages 121–124, June 1987.

[CK89]   P. K. Chan and K. Karplus. Computing signal delay in general RC networks by tree/link partitioning. In *Proceedings of the 26th ACM/IEEE Design Automation Conference*, pages 485–490, July 1989.

[CK91]   H. Y. Chen and S. M. Kang. iCOACH: A circuit optimization aid for CMOS high-performance circuits. *Integration, the VLSI Journal*, 10(2):185–212, January 1991.

[CK99]   P. Chen and K. Keutzer. Towards true crosstalk noise analysis. In *Proceedings of the IEEE/ACM International Conference on Computer-Aided Design*, pages 139–144, 1999.

[CK00]   Y. Cheng and S. M. Kang. A temperature-aware simulation environment for reliable ULSI chip design. *IEEE Transactions* on *Computer-Aided Design of Integrated Circuits and Systems*, 19(10):1211–1220, October 2000.

[CKK00]   P. Chen, D. A. Kirkpatrick, and K. Keutzer. Miller factor for gate-level coupling delay calculation. In *Proceedings of the IEEE/ACM International Conference on Computer-Aided Design*, pages 68–74, 2000.

[CL75]   L. O. Chua and P.-M. Lin. *Computer-aided analysis of electronic circuits: Algorithms and computational techniques*. Prentice-Hall, Englewood Cliffs, NJ, 1975.

[Cla61]   C. E. Clark. The greatest of a finite set of random variables. *Operational Research*, 9:85–91, 1961.

[CLR90]   T. H. Cormen, C. E. Leiserson, and R. L. Rivest. *Introduction to Algorithms*. McGraw-Hill, New York, NY, 1990.

[CMS01]   L. H. Chen and M. Marek-Sadowska. Aggressor alignment for worst-case crosstalk noise. *IEEE Transactions on Computer-Aided Design of Integrated Circuits and Systems*, 20(5):612–621, May 2001.

[CN94]   E. Chiprout and M. Nakhla. *Asymptotic Waveform Evaluation and Moment-Matching for Interconnect Analysis*. Kluwer Academic Publishers, Boston, MA, 1994.

[Coc02]   P. Cocchini. Concurrent flip-flop and repeater insertion for high-performance integrated circuits. In *Proceedings of the IEEE/ACM International Conference on Computer-Aided Design*, pages 268–273, 2002.

[CP95]   K. Chaudhary and M. Pedram. Computing the area versus delay trade-off curves in technology mapping. *IEEE Transactions on Computer-Aided Design of Integrated Circuits and Systems*, 14(12):1480–1489, December 1995.

[CPO02]   M. Celik, L. Pileggi, and A. Odabasioglu. *IC Interconnect Analysis*. Kluwer Academic Publishers, Boston, MA, 2002.

[CS89]   P. K. Chan and M. D. F. Schlag. Bounds on signal delay in RC mesh networks. *IEEE Transactions on Computer-Aided Design of Integrated Circuits and Systems*, 8(6):581–589, June 1989.

[CS92]   L. F. Chao and E. H.-M. Sha. Unfolding and retiming data-flow DSP programs for RISC multiprocessor scheduling. In *Proceedings of IEEE International Conference on Acoustics, Speech and Signal Processing*, pages 565–568, 1992.

[CS93a]   L. F. Chao and E. H.-M. Sha. Efficient retiming and unfolding. In *Proceedings of IEEE International Conference on Acoustics, Speech and Signal Processing*, pages 421–424, 1993.

[CS93b]   J. D. Cho and M. Sarrafzadeh. A buffer distribution algorithm for high-speed clock routing. In *Proceedings of the ACM/IEEE Design Automation Conference*, pages 537–543, 1993.

[CS96]   V. Chandramouli and K. A. Sakallah. Modeling the effects of temporal proximity of input transitions on gate propagation delay and transition time. In *Proceedings of the ACM/IEEE Design Automation Conference*, pages 617–622, 1996.

[CS03a]   H. Chang and S. S. Sapatnekar. Statistical timing analysis considering spatial correlations using a single PERT-like traversal. In *Proceedings of the IEEE/ACM International Conference on Computer-Aided Design*, pages 621–625, November 2003.

[CS03b]   G. Chen and S. S. Sapatnekar. Partition-driven standard cell thermal placement. In *Proceedings of the ACM International Symposium on Physical Design*, pages 75–80, 2003.

[CSH93]   W. Chuang, S. S. Sapatnekar, and I. N. Hajj. A unified algorithm for gate sizing and clock skew optimization to minimize sequential circuit area. In *Proceedings of the IEEE/ACM International Conference on Computer-Aided Design*, pages 220–223, 1993.

[CSH95]   W. Chuang, S. S. Sapatnekar, and I. N. Hajj. Timing and area optimization for standard cell VLSI circuit design. *IEEE Transactions on Computer-Aided Design of Integrated Circuits and Systems*, pages 308–320, March 1995.

[CW94]   Y. P. Chen and D. F. Wong. On retiming for FPGA logic module minimization. In *Proceedings of the IEEE International Conference on Computer Design*, pages 394 –397, 1994.

[CW96]   J. Cong and C. Wu. An improved algorithm for performance optimal technology mapping with retiming in LUT-based FPGA design. In *Proceedings of the IEEE International Conference on Computer Design*, pages 572–578, 1996.

[CW00]   B. Choi and D. M. H. Walker. Timing analysis of combinational circuits including capacitive coupling and statistical process variation. In *Proceedings of the IEEE VLSI Test Symposium*, pages 49–54, April 2000.

[DA89]   Z. Dai and K. Asada. MOSIZ : A two-step transistor sizing algorithm based on optimal timing assignment method for multi-stage complex gates. In *Proceedings of the 1989 Custom Integrated Circuits Conference*, pages 17.3.1–17.3.4, May 1989.

[Dai01]   W.-J. Dai. Hierarchical physical design methodology for multi-million gate chips. In *Proceedings of the ACM International Symposium on Physical Design*, pages 179–181, 2001.

[DB96]   D. K. Das and B. B. Bhattachaya. Does retiming affect redundancy in sequential circuits? In *Proceedings of the International Conference on VLSI Design*, pages 260–263, 1996.

[DBM03]   L. Ding, D. Blaauw, and P. Mazumder. Accurate crosstalk noise modeling for early signal integrity analysis. *IEEE Transactions on Computer-Aided Design of Integrated Circuits and Systems*, 22(5):627–634, May 2003.

[DC94]   S. Dey and S. Chakradhar. Retiming sequential circuits to enhance testability. In *Proceedings of the IEEE VLSI Test Symposium*, pages 28–33, 1994.

[DCJ96]   M. P. Desai, R. Cvijetic, and J. Jensen. Sizing of clock distribution networks for high performance CPU chips. In *Proceedings of the ACM/IEEE Design Automation Conference*, pages 389–394, 1996.

[De 94]   G. De Micheli. *Synthesis and Optimization of Digital Circuits*. McGraw-Hill, New York, NY, 1994.

[Deo94]   R. B. Deokar. Clock period minimization using skew optimization and retiming. Master's thesis, Department of Electrical and Computer Engineering, Iowa State University, Ames IA, May 1994.

[Dep95]   Department of Operations Research, Stanford University. *MINOS 5.4 USER'S GUIDE*, 1995.

[Dev97]   A. Devgan. Efficient coupled noise estimation for on-chip interconnects. In *Proceedings of the IEEE/ACM International Conference on Computer-Aided Design*, pages 147–153, 1997.

[DFHS89]   A. E. Dunlop, J. P. Fishburn, D. D. Hill, and D. D. Shugard. Experiments using automatic physical design techniques for optimizing circuit performance. In *Proceedings of the 32nd Midwest Symposium on Circuits and Systems, Urbana, IL*, August 1989.

[DH77]   S. W. Director and G. D. Hachtel. The simplicial approximation approach to design centering. *IEEE Transactions on Circuits and Systems*, CAS-24(7):363–372, July 1977.

[DIY91]   Y. Deguchi, N. Ishiura, and S. Yajima. Probabilistic ctss: Analysis of timing error probability in asynchronous logic circuits. In *Proceedings of the ACM/IEEE Design Automation Conference*, pages 650–655, June 1991.

[DK03]   A. Devgan and C. V. Kashyap. Block-based statistical timing analysis with uncertainty. In *Proceedings of the IEEE/ACM International Conference on Computer-Aided Design*, pages 607–614, November 2003.

[DMP96]   F. Dartu, N. Menezes, and L. T. Pileggi. Performance computation for precharacterized CMOS gates with RC loads. *IEEE Transactions on Computer-Aided Design of Integrated Circuits and Systems*, 15(5):544–553, May 1996.

[DMS⁺02]   A. Daga, L. Mize, S. Sripada, C. Wolff, and Q. Wu. Automated timing model generation. In *Proceedings of the ACM/IEEE Design Automation Conference*, pages 146 151, 2002.

[DP97]   F. Dartu and L. T. Pileggi. Computing worst-case gate delays due to dominant capacitance coupling. In *Proceedings of the ACM/IEEE Design Automation Conference*, pages 46–51, 1997.

[DPP95]   Y. G. DeCastelo-Vide-e-Souza, M. Potkonjak, and A. Parker. Optimal ILP-based approach for throughput optimization using simultaneous algorithm/architecure matching and retiming. In *Proceedings of the ACM/IEEE Design Automation Conference*, pages 113–118, 1995.

[DR69]   S. W. Director and R. A. Rohrer. The generalized adjoint network and network sensitivities. *IEEE Transactions on Circuit Theory*, 16(3):318–323, August 1969.

[DS94]   R. B. Deokar and S. S. Sapatnekar. A graph-theoretic approach to clock skew optimization. In *Proceedings of the IEEE International Symposium on Circuits and Systems*, pages 1.407–1.410, 1994.

[DS95]   R. B. Deokar and S. S. Sapatnekar. A fresh look at retiming via clock skew optimization. In *Proceedings of the ACM/IEEE Design Automation Conference*, pages 310–315, 1995.

[Dun92]   P. Duncan *et al.* Hi-PASS: A computer-aided synthesis system for maximally parallel digital signal processing ASICS. In *Proceedings of IEEE International Conference on Acoustics, Speech and Signal Processing*, pages V–605–608, 1992.

[DY96]   M. P. Desai and Y.-T. Yen. A systematic technique for verifying critical path delays in a 300MHz Alpha CPU design using circuit simulation. In *Proceedings of the ACM/IEEE Design Automation Conference*, pages 125–130, 1996.

[Eck80]   J. G. Ecker. Geometric programming: methods, computations and applications. *SIAM Review*, 22(3):338–362, July 1980.

[Eda91]   M. Edahiro. Minimum skew and minimum path length routing in VLSI layout design. *NEC Research and Development*, 32:569–575, 1991.

[Eda93]   M. Edahiro. Delay minimization for zero skew routing. In *Proceedings of the IEEE/ACM International Conference on Computer-Aided Design*, pages 563–566, 1993.

[Edw03]   S. A. Edwards. Making cyclic circuits acyclic. In *Proceedings of the ACM/IEEE Design Automation Conference*, pages 159–162, 2003.

[EL97]   I. M. Elfadel and D. D. Ling. A block rational Arnoldi algorithm for multipoint passive model-order reduction of multiport RLC networks. In *Proceedings of the IEEE/ACM International Conference on Computer-Aided Design*, pages 66–71, 1997.

[Elm48]   W. C. Elmore. The transient response of damped linear networks with particular regard to wideband amplifiers. *Journal of Applied Physics*, 19, January 1948.

[EMRM95]   A. El-Maleh, T. Marchok, J. Rajski, and W. Maly. On test set preservation of retimed circuits. In *Proceedings of the ACM/IEEE Design Automation Conference*, pages 176–182, 1995.

[EMRM97]   A. El-Maleh, T. Marchok, J. Rajski, and W. Maly. Behavior and testability preservation under the retiming transformation. *IEEE Transactions on Computer-Aided Design of Integrated Circuits and Systems*, 16(5):528–543, May 1997.

[ESS96]   G. Even, I. Y. Spillinger, and L. Stok. Retiming revisited and reversed. *IEEE Transactions on Computer-Aided Design of Integrated Circuits and Systems*, 15(3):348–357, March 1996.

[FD85]   J.P. Fishburn and A.E. Dunlop. TILOS: A posynomial programming approach to transistor sizing. In *Proceedings of the IEEE International Conference on Computer-Aided Design*, pages 326–328, 1985.

[FF95]   P. Feldmann and R. W. Freund. Efficient linear circuit analysis by Padé approximation via the Lanczos process. *IEEE Transactions on Computer-Aided Design of Integrated Circuits and Systems*, 14(5):639–649, May 1995.

[FF96]   R. W. Freund and P. Feldmann. Reduced order modeling of large passive linear circuits by means of the SyPVL algorithm. In *Proceedings of the IEEE/ACM International Conference on Computer-Aided Design*, pages 280–287, 1996.

[FF98]   R. W. Freund and P. Feldmann. Reduced-order modeling of large
            linear passive multi-terminal circuits using matrix-Padé approxi-
            mation. In *Proceedings of Design, Automation and Test in Europe
            Conference*, pages 530–537, 1998.

[Fis90]   J. P. Fishburn. Clock skew optimization. *IEEE Transactions on
            Computers*, 39(7):945–951, July 1990.

[FK82]   A. L. Fisher and H. T. Kung. Synchronizing large systolic arrays.
            In *Proceedings of the SPIE*, volume 341, pages 44–52, May 1982.

[Fre99]   R. W. Freund. Passive reduced order models for interconnect
            simulation and their computation via Krylov subspace algorithms.
            In *Proceedings of the ACM/IEEE Design Automation Conference*,
            pages 195–200, 1999.

[Fri95]   E. G. Friedman, editor. *Clock distribution networks in VLSI cir-
            cuits and systems*. IEEE Press, New York, NY, 1995.

[GARP98]   P. D. Gross, R. Arunachalam, K. Rajagopal, and L. T. Pileggi.
            Determination of worst-case aggressor alignment for delay calcula-
            tion. In *Proceedings of the IEEE/ACM International Conference
            on Computer-Aided Design*, pages 212–219, 1998.

[GBW+01]   K. Gala, D. Blaauw, J. Wang, V. Zolotov, and M. Zhao. In-
            ductance 101: Analysis and design issues. In *Proceedings of the
            ACM/IEEE Design Automation Conference*, pages 329–334, 2001.

[Geb93]   C. H. Gebotys. Throughput optimized architectural synthesis.
            IEEE Transactions on VLSI Systems, 1(3):254–261, 1993.

[GLC94]   C. T. Gray, W. Liu, and R. K. Cavin, III. *Wave Pipelining: The-
            ory and CMOS Implementation*. Kluwer Academic Publishers,
            Boston, MA, 1994.

[GS99]   E. Goldberg and A. Saldanha. Timing analysis with implicitly
            specified false paths. In *Workshop Notes, International Workshop
            on Timing Issues in the Specification and Synthesis of Digital Sys-
            tems*, pages 157–164, 1999.

[GS03]   B. Goplen and S. S. Sapatnekar. Efficient thermal placement
            of standard cells in 3d ic?s using a force directed approach.
            In *Proceedings of the IEEE/ACM International Conference on
            Computer-Aided Design*, pages 86–89, 2003.

[GTP97]   R. Gupta, B. Tutuianu, and L. T. Pileggi. The Elmore delay as a
            bound for RC trees with generalized input signals. *IEEE Transac-
            tions on Computer-Aided Design of Integrated Circuits and Sys-
            tems*, 16(1):95–104, January 1997.

[Gv96]   G. H. Golub and C. F. van Loan. *Matrix computations*. Johns Hopkins University Press, 3rd edition, 1996.

[Has97]   S. Hassoun. *Architectural Retiming: A Technique for Optimizing Latency-Constrained Circuits*. PhD thesis, Department of Computer Science and Engineering, University of Washington, Seattle, WA, 1997.

[HAT02]   S. Hassoun, C. J. Alpert, and M. Thiagarajan. Optimal buffered routing path constructions for single and multiple clock domain systems. In *Proceedings of the IEEE/ACM International Conference on Computer-Aided Design*, pages 247–253, 2002.

[HCC96]   S. Y. Huang, K. T. Cheng, and K. C. Chen. On verifying the correctness of retimed circuits. In *Proceedings of the Great Lakes Symposium on VLSI*, pages 277–280, 1996.

[HCC97]   S. Y. Huang, K. T. Cheng, and K. C. Chen. AQUILA: An equivalence verifier for large sequential circuits. In *Proceedings of the Asia/South Pacific Design Automation Conference*, pages 455–460, 1997.

[Hed87]   K. S. Hedlund. AESOP : A tool for automated transistor sizing. In *Proceedings of the 24th ACM/IEEE Design Automation Conference*, pages 114–120, June 1987.

[HF91]   L. S. Heusler and W. Fichtner. Transistor sizing for large combinational digital CMOS circuits. *Integration, the VLSI journal*, 10(2):155–168, January 1991.

[HH97]   D. Harris and M. A. Horowitz. Skew-tolerant domino circuits. *IEEE Journal of Solid-State Circuits*, 32(11):1702–1711, November 1997.

[Hit82]   R. B. Hitchcock, Sr. Timing verification and the timing analysis program. In *Proceedings of the 19th Design Automation Conference*, pages 594–604, 1982.

[HJ87]   N. Hedenstierna and K. O. Jeppson. CMOS circuit speed and buffer optimization. *IEEE Transactions on Computer-Aided Design of Integrated Circuits and Systems*, CAD-6:270–281, March 1987.

[HKK95]   Y. Higami, S. Kajihara, and K. Kinoshita. Test sequence compaction by reduced scan shift and retiming. In *Proceedings of the IEEE Asian Test Symposium*, pages 169–175, 1995.

[HKK96]   Y. Higami, S. Kajihara, and K. Kinoshita. Partially parallel scan chain for test length reduction by using retiming technique. In

Proceedings of the IEEE Asian Test Symposium, pages 94–99, 1996.

[HLFC97]   H.-Y. Hsieh, W. Liu, P. Franzon, and R. Cavin III.   Clocking optimization and distribution in digital systems with scheduled skews. *Journal of VLSI Signal Processing Systems for Signal, Image and Video Technology*, 16(2/3):131–148, June/July 1997.

[HN01]   D. Harris and S. Naffziger. Statistical cock skew modeling with data delay variations. *IEEE Transactions on VLSI Systems*, 9(6):888–898, December 2001.

[HS00]   J. Hu and S. S. Sapatnekar. Algorithms for non-hanan-based optimization for VLSI interconnect under a higher order AWE model. *IEEE Transactions on Computer-Aided Design of Integrated Circuits and Systems*, 19(4):446–458, April 2000.

[HSFK89]   D. Hill, D. Shugard, J. Fishburn, and K. Keutzer. *Algorithms and Techniques for VLSI Layout Synthesis.* Kluwer Academic Publishers, Boston, MA, 1989.

[HW96]   M. Heshami and B. A. Wooley. A 250-MHz skewed-clock pipelined data buffer. *IEEE Journal of Solid-State Circuits*, 31(3):376–383, March 1996.

[HWMS03]   B. Hu, Y. Watanabe, and M. Marek-Sadowska. Gain-based technology mapping for discrete-size cell libraries. In *Proceedings of the ACM/IEEE Design Automation Conference*, pages 574–579, 2003.

[ILP92]   A. Ishii, C. E. Leiserson, and M. C. Papaefthymiou. Optimizing two-phase, level-clocked circuitry. In *Advanced Research in VLSI and Parallel Systems: Proceedings of the 1992 Brown/MIT Conference*, pages 246–264, 1992.

[IP96]   A. T. Ishii and M. C. Papaefthymiou. Efficient pipelining of level-clocked circuits with min-max propagation delays. In *Workshop Notes, International Workshop on Timing Issues in the Specification and Synthesis of Digital Systems*, pages 39–51, 1996.

[Ish93]   A. T. Ishii. Retiming gated-clocks and precharged circuit structures. In *Proceedings of the IEEE/ACM International Conference on Computer-Aided Design*, pages 300–307, 1993.

[JB00]   E. Jacobs and M. R. C. M. Berkelaar. Gate sizing using a statistical delay model. In *Proceedings of Design, Automation and Test in Europe Conference*, pages 283–290, 2000.

[JC93]   D. Joy and M. Ciesielski. Clock period minimization with wave pipelining. *IEEE Transactions on Computer-Aided Design of Integrated Circuits and Systems*, 12(4):461–472, April 1993.

[JH01]   X. Jiang and S. Horiguchi. Statistical skew modeling for general clock distribution networks in presence of process variations. *IEEE Transactions on VLSI Systems*, 9(5):704–717, October 2001.

[JKN+03]   J. A. G. Jess, K. Kalafala, S. R. Naiddu, R. H. J. M. Otten, and C. Visweswariah. Statistical timing for parametric yield prediction of digital integrated circuits. In *Proceedings of the ACM/IEEE Design Automation Conference*, pages 932–937, June 2003.

[JMDK93]   H.F. Jyu, S. Malik, S. Devadas, and K.W. Keutzer. Statistical timing analysis of combinational logic circuits. *IEEE Transactions on VLSI Systems*, 1(2):126–137, June 1993.

[Jou87a]   N. Jouppi. Derivation of signal flow direction in MOS VLSI. *IEEE Transactions on Computer-Aided Design of Integrated Circuits and Systems*, CAD-6(5):480–490, May 1987.

[Jou87b]   N. P. Jouppi. Timing analysis and performance improvement of MOS VLSI design. *IEEE Transactions on Computer-Aided Design of Integrated Circuits and Systems*, CAD-4(4):650–665, July 1987.

[JS91]   Yun-Cheng Ju and Resve A. Saleh. Incremental techniques for the identification of statically sensitizable critical paths. In *Proceedings of the ACM/IEEE Design Automation Conference*, pages 541–546, 1991.

[JSK90]   M. A. B. Jackson, A. Srinivasan, and E. S. Kuh. Clock routing for high-performance IC's. In *Proceedings of the ACM/IEEE Design Automation Conference*, pages 573–579, 1990.

[KBD+01]   N. A. Kurd, J. S. Barkatullah, R. O. Dizon, T. D. Fletcher, and P. D. Madland. A multigigahertz clocking scheme for the Pentium® 4 microprocessor. *IEEE Journal of Solid-State Circuits*, 36(11):1647–1653, November 2001.

[KC66]   T. I. Kirkpatrick and N. R. Clark. PERT as an aid to logic design. *IBM Journal of Research and Development*, 10(2):135–141, March 1966.

[KCR91]   A. Kahng, J. Cong, and G. Robins. High-performance clock routing based on recursive geometric matching. In *Proceedings of the ACM/IEEE Design Automation Conference*, pages 322–327, 1991.

[Keu87]   K. Keutzer. DAGON: Technology binding and local optimization by DAG matching. In *Proceedings of the ACM/IEEE Design Automation Conference*, pages 341–347, 1987.

[KGV83]   S. Kirkpatrick, C. D. Gelatt, Jr., and M. P. Vecchi. Optimization by simulated annealing. *Science*, 220(4598):671–680, May 1983.

[KK00]   C. V. Kashyap and B. L. Krauter. A realizable driving point model for on-chip interconnect with inductance. In *Proceedings of the ACM/IEEE Design Automation Conference*, pages 190–195, 2000.

[KKS00]   K. Kasamsetty, M. Ketkar, and S. S. Sapatnekar. A new class of convex functions for delay modeling and their application to the transistor sizing problem. *IEEE Transactions on Computer-Aided Design of Integrated Circuits and Systems*, 19(7):779–788, July 2000.

[KM98]   A. Kahng and S. Muddu. New efficient algorithms for computing effective capacitance. In *Proceedings of the ACM International Symposium on Physical Design*, pages 147–151, 1998.

[KMS99]   A. B. Kahng, S. Muddu, and E. Sarto. Tuning strategies for global interconnects in high-performance deep submicron ic's. *VLSI Design*, 10(1):21–34, 1999.

[KO95]   I. Karkowski and R. H. J. M. Otten. Retiming synchronous circuitry with imprecise delays. In *Proceedings of the ACM/IEEE Design Automation Conference*, pages 322–326, 1995.

[KP98]   R. Kay and L. Pileggi. PRIMO: Probability interpretation of moments for delay calculation. In *Proceedings of the ACM/IEEE Design Automation Conference*, pages 463–468, 1998.

[KR95]   A. B. Kahng and G Robins. *On Optimal Interconnections for VLSI*. Kluwer Academic Publishers, Boston, MA, 1995.

[KS01]   M. Kuhlmann and S. S. Sapatnekar. Exact and efficient crosstalk estimation. *IEEE Transactions on Computer-Aided Design of Integrated Circuits and Systems*, 20(7):858 – 866, July 2001.

[KS02]   M. Ketkar and S. S. Sapatnekar. Standby power optimization via transistor sizing and dual threshold voltage optimization. In *Proceedings of the IEEE/ACM International Conference on Computer-Aided Design*, pages 375–378, 2002.

[KSV96]   D. A. Kirkpatrick and A. L. Sangiovanni-Vincentelli. Digital sensitivity: Predicting signal interaction using functional analysis. In *Proceedings of the IEEE/ACM International Conference on Computer-Aided Design*, pages 536–541, 1996.

[KT94]   D. Kagaris and S. Tragoudas. Retiming algorithm with application to VLSI testability. In *Proceedings of the Great Lakes Symposium on VLSI*, pages 216–221, 1994.

[KTB93]   D. Kagaris, S. Tragoudas, and D. Bhatia. Pseudo-exhaustive BIST for sequential circuits. In *Proceedings of the IEEE International Conference on Computer Design*, pages 523–527, 1993.

[Kuo93]   B. C. Kuo. *Automatic Control Systems*. Prentice-Hall, Englewood Cliffs, NJ, 6th edition, 1993.

[KY98]   K. J. Kerns and A. T. Yang. Preservation of passivity during RLC network reduction via split congruence transformations. *IEEE Transactions on Computer-Aided Design of Integrated Circuits and Systems*, 17(7):582–591, July 1998.

[LAP98]   T. Lin, E. Acar, and L. Pileggi. h-gamma: An RC delay metric based on a gamma distribution approximation of the homogeneous response. In *Proceedings of the IEEE/ACM International Conference on Computer-Aided Design*, pages 19–25, 1998.

[LBB+00]   R. Levy, D. Blaauw, G. Braca, A. Dasgupta, A. Grinshpon, C. Oh, B. Orshav, S. Sirichotiyakul, and V. Zolotov. ClariNet: A noise analysis tool for deep submicron design. In *Proceedings of the ACM/IEEE Design Automation Conference*, pages 233–238, 2000.

[LCKK01]   J. J. Liou, K. T. Cheng, S. Kundu, and A. Krstic. Fast statistical timing analysis by probabilistic event propagation. In *Proceedings of the ACM/IEEE Design Automation Conference*, pages 661–666, June 2001.

[LE91]   B. Lockyear and C. Ebeling. Optimal retiming of level-clocked circuits using symmetric clock schedules. Technical Report UW-CSE-91-10-01, Department of Computer Science and Engineering, University of Washington, Seattle, 1991.

[LE92]   B. Lockyear and C. Ebeling. Optimal retiming of multi-phase level-clocked circuits. In *Advanced Research in VLSI and Parallel Systems: Procedings of the 1992 Brown/MIT Conference*, pages 265–280, 1992.

[LE94]   B. Lockyear and C. Ebeling. Optimal retiming of level-clocked circuits using symmetric clock schedules. *IEEE Transactions on Computer-Aided Design of Integrated Circuits and Systems*, 13(9):1097–1109, September 1994.

[Lin93]   B. Lin. Restructuring of synchronous logic circuits. In *Proceedings of the European Design Automation Conference*, pages 205–209, 1993.

[LKA95a]   S. Lejmi, B. Kaminska, and B. Ayari. Retiming for BIST-sequential circuits. In *Proceedings of the IEEE International Symposium on Circuits and Systems*, pages 1740–1742, 1995.

[LKA95b]   S. Lejmi, B. Kaminska, and B. Ayari. Retiming, resynthesis, and partitioning for the pseudo-exhaustive testing of sequential circuits. In *Proceedings of the IEEE VLSI Test Symposium*, pages 434–439, 1995.

[LKA95c]   S. Lejmi, B. Kaminska, and B. Ayari. Synthesis and retiming for the pseudo-exhaustive BIST of synchronous sequential circuits. In *Proceedings of the IEEE International Test Conference*, pages 683–692, 1995.

[LKA02]   F. Liu, C. V. Kashyap, and C. J. Alpert. A delay metric for rc circuits based on the weibull distribution. In *Proceedings of the IEEE/ACM International Conference on Computer-Aided Design*, pages 620–624, 2002.

[LKW93]   S. Lejmi, B. Kaminska, and E. Wagneur. Resynthesis and retiming of synchronous sequential circuits. In *Proceedings of the IEEE International Symposium on Circuits and Systems*, pages 1674–1677, 1993.

[LKWC02]   J. J. Liou, A. Krstic, L. C. Wang, and K. T. Cheng. False-path-aware statistical timing analysis and efficient path selection for delay testing and timing validation. In *Proceedings of the ACM/IEEE Design Automation Conference*, pages 566–569, June 2002.

[LM84]   T.-M. Lin and C. A. Mead. Signal delay in general RC networks. *IEEE Transactions on Computer-Aided Design of Integrated Circuits and Systems*, 3(4), October 1984.

[LNPS00]   Y. Liu, S. R. Nassif, L. T. Pileggi, and A. J. Strojwas. Impact of interconnect variations on the clock skew of a gigahertz microprocessor. In *Proceedings of the ACM/IEEE Design Automation Conference*, pages 168–171, June 2000.

[LP95a]   K. N. Lalgudi and M. Papaefthymiou. DELAY: An efficient tool for retiming with realistic delay modeling. In *Proceedings of the ACM/IEEE Design Automation Conference*, pages 304–309, 1995.

[LP95b]   K. N. Lalgudi and M. C. Papaefthymiou. Efficient retiming under a general delay model. In *Advanced Research in VLSI : the 1995 MIT/UNC-Chapel Hill Conference*, pages 368–382, 1995.

[LP96]   K. N. Lalgudi and M. Papaefthymiou. Fixed-phase retiming for low power. In *Proceedings of the International Symposium of Low Power Electronics and Design*, pages 259–264, 1996.

[LRS83]   C. Leiserson, F. Rose, and J. B. Saxe. Optimizing synchronous circuitry by retiming. In *Proceedings of the 3rd Caltech Conference on VLSI*, pages 87–116, 1983.

[LRSV83]   E. Leelarasmee, A. E. Ruehli, and A. L. Sangiovanni-Vincentelli. The waveform relaxation method for the time-domain analysis of large scale integrated circuits and systems. *IEEE Transactions on Computer-Aided Design of Integrated Circuits and Systems*, CAD-1(3):131–145, July 1983.

[LS91]   C. E. Leiserson and J. B. Saxe. Retiming synchronous circuitry. *Algorithmica*, 6:5–35, 1991.

[LS96]   D. Lehther and S. S. Sapatnekar. Clock tree synthesis for multichip modules. In *Proceedings of the IEEE/ACM International Conference on Computer-Aided Design*, pages 53–56, 1996.

[LS98]   D. Lehther and S. S. Sapatnekar. Moment-based techniques for rlc clock tree construction. *IEEE Transactions on Circuits and Systems II: Analog and Digital Signal Processing*, 45(1):69–79, January 1998.

[LSC+93]   L.-T. Liu, M. Shih, N.-C. Chou, C.-K. Cheng, and W. Ku. Performance-driven partitioning using retiming and replication. In *Proceedings of the IEEE/ACM International Conference on Computer-Aided Design*, pages 296–299, 1993.

[LTW94]   J.-f. Lee, D. T. Tang, and C. K. Wong. A timing analysis algorithm for circuits with level-sensitive latches. In *Proceedings of the IEEE/ACM International Conference on Computer-Aided Design*, pages 743–748, 1994.

[Lue84]   D. G. Luenberger. *Linear and Nonlinear Programming*. Addison-Wesley, Reading, Massachusetts, 1984.

[LVW97]   C. Legl, P. Vanbekbergen, and A. Wang. Retiming of edge-triggered circuits with multiple clocks and load enables. In *Workshop Notes, International Workshop on Logic Synthesis*, 1997.

[LW83]   Y. Z. Liao and C. K. Wong. An algorithm to compact a VLSI symbolic layout with mixed constraints. *IEEE Transactions on Computer-Aided Design of Integrated Circuits and Systems*, 2(2):62–69, April 1983.

[LZ03]   C. Lin and H. Zhou. Retiming for wire pipelining in system-on-chip. In *Proceedings of the IEEE/ACM International Conference on Computer-Aided Design*, pages 215–220, 2003.

[Mal95]   S. Malik. Analysis of cyclic combinational circuits. *IEEE Transactions on Computer-Aided Design of Integrated Circuits and Systems*, 13(7):950–956, July 1995.

[Mar86]   D. Marple. Performance optimization of digital VLSI circuits.
           Technical Report CSL-TR-86-308, Stanford University, October
           1986.

[Mar96]   H.-G. Martin. Retiming for circuits with enable registers. In
           Proceedings of EUROMICRO-22, pages 275–280, 1996.

[MB91]   P. C. McGeer and R. K. Brayton. *Integrating Functional and
           Temporal Domains in Logic Design*. Kluwer Academic Publishers,
           Boston, MA, 1991.

[MDG93]   J. Monteiro, S. Devadas, and A. Ghosh. Retiming sequential
           circuits for low power. In *Proceedings of the IEEE/ACM Inter-
           national Conference on Computer-Aided Design*, pages 398–402,
           1993.

[MEMR96]   T. Marchok, A. El-Maleh, W. Maly, and J. Rajski. A complexity
           analysis of sequential ATPG. *IEEE Transactions on Computer-
           Aided Design of Integrated Circuits and Systems*, 15(11):1409–
           1423, November 1996.

[MG86]   D. Marple and A. El Gamal. Area-delay optimization of pro-
           grammable logic arrays. In *Fourth MIT Conference on VLSI*,
           pages 171–194, April 1986.

[MG87]   D. Marple and A. El Gamal. Optimal selection of transistor sizes
           in digital VLSI circuits. In *Stanford Conference on VLSI*, pages
           151–172, 1987.

[MH91]   A. Münzner and G Hemme. Converting combinational circuits
           into pipelined data paths. In *Proceedings of the IEEE/ACM Inter-
           national Conference on Computer-Aided Design*, pages 368–371,
           1991.

[Mil28]   A. A. Milne. *The House at Pooh Corner*. E. P. Dutton and Co.,
           New York, NY, 1928.

[Mor76]   D. F. Morrison. *Multivariate Statistical Methods*. McGraw-Hill,
           New York, NY, 1976.

[MS96]   N. Maheshwari and S. S. Sapatnekar. A practical algorithm for
           retiming level-clocked circuits. In *Proceedings of the IEEE Inter-
           national Conference on Computer Design*, pages 440–445, 1996.

[MS97a]   N. Maheshwari and S. S. Sapatnekar. An improved algorithm for
           minimum-area retiming. In *Proceedings of the ACM/IEEE Design
           Automation Conference*, pages 2–7, 1997.

[MS97b]   N. Maheshwari and S. S. Sapatnekar. Minimum area retiming
           with equivalent initial states. In *Proceedings of the IEEE/ACM*

International Conference on Computer-Aided Design, pages 216–219, 1997.

[MS98a]   N. Maheshwari and S. S. Sapatnekar. Efficient minarea retiming for large level-clocked circuits. In Proceedings of the Conference on Design Automation and Test in Europe, pages 840–845, 1998.

[MS98b]   N. Maheshwari and S. S. Sapatnekar. Efficient retiming of large circuits. IEEE Transactions on VLSI Systems, 6(1):74–83, March 1998.

[MS99]   N. Maheshwari and S. S. Sapatnekar. Optimizing large multiphase level-clocked circuits. IEEE Transactions on Computer-Aided Design of Integrated Circuits and Systems, 18(9):1249–1264, September 1999.

[MSBS90]   S. Malik, E. M. Sentovich, R. K. Brayton, and A. Sangiovanni-Vincentelli. Retiming and resynthesis: Optimizing sequential networks with combinational techniques. In Proceedings of the 23rd Anual Hawaii International Conference on System Sciences, pages 397–406, 1990.

[MSBS91]   S. Malik, E. M. Sentovich, R. K. Brayton, and A. Sangiovanni-Vincentelli. Retiming and resynthesis: Optimizing sequential networks with combinational techniques. IEEE Transactions on Computer-Aided Design of Integrated Circuits and Systems, 10(1):74–84, January 1991.

[MSM04]   M. Mneimneh, K. A. Sakallah, and J. Moondanos. Preserving synchronizing sequences of sequential circuits after retiming. In Proceedings of the Asia/South Pacific Design Automation Conference, 2004.

[Nag75]   L. W. Nagel. Spice2: A computer program to simulate semiconductor circuits. Technical Report ERL M520, Electronics Research Laboratory, University of California, Berkeley, Berkeley, CA, May 1975.

[Nai02]   S. Naidu. Timing yield calculation using an impulse-train approach. In Proceedings of 15th International Conference on VLSI Design, pages 219–224, January 2002.

[NBHY89]   R. Nair, C. L. Berman, P. S. Hauge, and E. J. Yoffa. Generation of performance constraints for layout. IEEE Transactions on Computer-Aided Design of Integrated Circuits and Systems, 8(8):860–874, August 1989.

[NF96]   J. L. Neves and E. G. Friedman. Design methodology for synthesizing clock distribution networks exploiting nonzero localized clock skew. *IEEE Transactions on VLSI Systems*, 4(2):286–291, June 1996.

[NF97]   J. L. Neves and E. G. Friedman. Buffered clock tree synthesis with non-zero clock skew scheduling for increased tolerance to process parameter variation. *Journal of VLSI Signal Processing Systems for Signal, Image and Video Technology*, 16(2/3):149–162, June/July 1997.

[NK01]   I. Neumann and W. Kunz. Placement driven retiming with a coupled edge timing model. In *Proceedings of the IEEE/ACM International Conference on Computer-Aided Design*, pages 95–102, 2001.

[NR00]   J. W. Nilsson and S. A. Riedel. *Electric Circuits*. Prentice-Hall, Upper Saddle River, NJ, 6th edition, 2000.

[NS04]   V. Nookala and S. S. Sapatnekar. A method for correcting the functionality of a wire-pipelined circuit. In *Proceedings of the ACM/IEEE Design Automation Conference*, 2004.

[OCP98]   A. Odabasioglu, M. Celik, and L. T. Pileggi. PRIMA: Passive reduced-order interconnect macromodeling algorithm. *IEEE Transactions on Computer-Aided Design of Integrated Circuits and Systems*, 17(8):645–654, August 1998.

[OCP99]   A. Odabasioglu, M. Celik, and L. T. Pileggi. Practical considerations for passive reduction of RLC circuits. In *Proceedings of the IEEE/ACM International Conference on Computer-Aided Design*, pages 214–219, 1999.

[OK02]   M. Orshansky and K. Keutzer. A general probabilistic framework for worst case timing analysis. In *Proceedings of the ACM/IEEE Design Automation Conference*, pages 556–561, June 2002.

[OMC⁺02] M. Orshansky, L. Milor, P. Chen, K. Keutzer, and C. Hu. Impact of spatial intrachip gate length variability on the performance of high-speed digital circuits. *IEEE Transactions on Computer-Aided Design of Integrated Circuits and Systems*, 21(5):544–553, May 2002.

[OS89]   P. R. O'Brien and D. T. Savarino. Modeling the driving-point characteristic of resistive interconnect for accurate delay estimation. In *Proceedings of the IEEE/ACM International Conference on Computer-Aided Design*, pages 512–515, 1989.

[OYO03]   K. Okada, K. Yamaoka, and H. Onodera. A statistical gate-delay model considering intra-gate variability. In *Proceedings of* the *IEEE/ACM International Conference on Computer-Aided Design*, pages 908–913, November 2003.

[Pan97]   P. Pan. Continuous retiming: Algorithms and applications. In *Proceedings of the IEEE International Conference on Computer Design*, pages 116–121, 1997.

[Pap91]   A. Papoulis. *Probability, Random Variables, and Stochastic Processes*. McGraw-Hill, New York, NY, 3rd edition, 1991.

[Pap93]   M. C. Papaefthymiou. *A Timing Analysis and Optimzation System for Level-Clocked Circuitry*. PhD thesis, Massachusetts Institute of Technology, Cambridge, MA, 1993.

[Pap98]   M. C. Papaefthymiou. Asymptotically efficient retiming under setup and hold constraints. In *Proceedings of the IEEE/ACM International Conference on Computer-Aided Design*, pages 288–295, 1998.

[PDS03]   J. R. Phillips, L. Daniel, and L. M. Silveira. Guaranteed passive balancing transformations for model order reduction. *IEEE Transactions on Computer-Aided Design of Integrated Circuits and Systems*, 22(8):1027–1041, August 2003.

[PL95]   P. Pan and C. L. Liu. Optimal clock period technology mapping for FPGA circuits. In *Proceedings of the ACM/IEEE Design Automation Conference*, pages 720–725, 1995.

[PMOP93]   S. Pullela, N. Menezes, J. Omar, and L. T. Pillage. Skew and delay optimization for reliable buffered clock trees. In *Proceedings of the IEEE/ACM International Conference on Computer-Aided Design*, pages 556–562, 1993.

[PMP93]   S. Pullela, N. Menezes, and L. T. Pillage. Reliable nonzero clock skew trees using wire width optimization. In *Proceedings of the ACM/IEEE Design Automation Conference*, pages 165–170, 1993.

[PR90]   L. T. Pillage and R. A. Rohrer. Asymptotic waveform evaluation for timing analysis. *IEEE Transactions on Computer-Aided Design of Integrated Circuits and Systems*, 9(4):352–366, April 1990.

[PR91]   M. Potkonjak and J. Rabaey. Optimizing resource utilization using transformations. In *Proceedings of the IEEE/ACM International Conference on Computer-Aided Design*, pages 88–91, 1991.

[PR92]   M. Potkonjak and J. Rabaey. Fast implementation of recursive programs using transformations. In *Proceedings of IEEE International Conference on Acoustics, Speech and Signal Processing*, pages 304–308, 1992.

[PR93a]   M. C. Papaefthymiou and K. H. Randall. Edge-triggered vs. two-phase level-clocking. In *Research on Integrated Systems: Proceedings of the 1993 Symposium*, March 1993.

[PR93b]   M. C. Papaefthymiou and K. H. Randall. TIM: A timing package for two-phase, level-clocked circuitry. In *Proceedings of the ACM/IEEE Design Automation Conference*, pages 497–502, 1993.

[PRV95]   L. T. Pillage, R. A. Rohrer, and C. Visweswariah. *Electronic circuit and system simulation methods*. McGraw-Hill, New York, NY, 1995.

[PS96]   N. L. Passos and E. H.-M. Sha. Achieving full parallelism using multidimensional retiming. *IEEE Transactions on Parallel and Dsitributed Systems*, 7(11):1150–1163, November 1996.

[PSB96]   N. L. Passos, E. H.-M. Sha, and S. C. Bass. Optimizing DSP flow graphs via schedule-based multidimensional retiming. *IEEE Transactions on Signal Processing*, 44(1):150–155, January 1996.

[QC03]   Z. Qin and C.-K. Cheng. Realizable parasitic reduction using generalized Y-Δ transformation. In *Proceedings of the ACM/IEEE Design Automation Conference*, pages 220–225, 2003.

[QPP94]   J. Qian, S. Pullela, and L. T. Pillage. Modeling the "effective capacitance" for the RC interconnect of CMOS gates. *IEEE Transactions on Computer-Aided Design of Integrated Circuits and Systems*, 13(12):1526–1535, December 1994.

[RB03]   M. D. Riedel and J. Bruck. The synthesis of cyclic combinational circuits. In *Proceedings of the ACM/IEEE Design Automation Conference*, pages 163–168, 2003.

[RCE+02]   P. J. Restle, C. A. Carter, J. P. Eckhardt, B. L. Krauter, B. D. McCredie, K. A. Jenkins, A. J. Weger, and A. V. Mule. The clock distribution of the Power4 microprocessor. In *Proceedings of the IEEE International Solid-State Circuits Conference*, pages 144–145, 2002.

[Ric66]   M. Richardson. *College Algebra*. Prentice-Hall, Englewood Cliffs, NJ, 3rd edition, 1966.

[RMW+01]   P. J. Restle, T. G. McNamara, D. A. Webber, P. J. Camporese, K. F. Eng, K. A. Jenkins, D. H. Allen, M. J. Rohn, M. P. Quaranta, D. W. Boerstler, C. J. Alpert, C. A. Carter, R. N. Bailey, J. G. Petrovick, B. L. Krauter, and B. D. McCredie. A clock distribution network for microprocessors. *IEEE Journal of Solid-State Circuits*, 36(5):792–799, May 2001.

[RPH83]   J. Rubenstein, P. Penfield, and M. A. Horowitz. Signal delay in RC tree networks. *IEEE Transactions on Computer-Aided Design of Integrated Circuits and Systems*, CAD-2(3):202–211, July 1983.

[RSSB98]   R. K. Ranjan, V. Singhal, F. Somenzi, and R. K. Brayton. Using combinational verification for sequential circuits. In *Workshop Notes, International Workshop on Logic Synthesis*, pages 290–298, 1998.

[Sap96]   S. S. Sapatnekar. Efficient calculation of all-pair input-to-output delays in synchronous sequential circuits. In *Proceedings of the IEEE International Symposium on Circuits and Systems*, pages IV.520–IV.523, 1996.

[Sap99]   S. S. Sapatnekar. On the chicken-and-egg problem of determining the effect of crosstalk on delay in integrated circuits. In *Proceedings of the IEEE 8th Topical Meeting on Electrical Performance of Electronic Packaging (EPEP-99)*, pages 245–248, 1999.

[Sap00]   S. S. Sapatnekar. A timing model incorporating the effect of crosstalk on delay and its application to optimal channel routing. *IEEE Transactions on Computer-Aided Design of Integrated Circuits and Systems*, 19(5):550–559, May 2000.

[Sax85]   J. B. Saxe. *Decomposable Searching Problems and Circuit Optimization by Retiming: Two Studies in General Transformations of Computational Structures*. PhD thesis, Carnegie-Mellon University, Pittsburgh, PA, 1985.

[SBS91]   N. Shenoy, R. K. Brayton, and A. Sangiovanni-Vincentelli. Retiming of circuits with single phase transparent latches. In *Proceedings of the IEEE International Conference on Computer Design*, pages 86–89, 1991.

[SBS92]   N. Shenoy, R. K. Brayton, and A. Sangiovanni-Vincentelli. Graph algorithms for clock schedule optimization. In *Proceedings of the IEEE/ACM International Conference on Computer-Aided Design*, pages 132–136, 1992.

[SBS93]   N. V. Shenoy, R. K. Brayton, and A. L. Sangiovanni-Vincentelli. Minimum padding to satisfy short path constraints. In *Proceedings of the IEEE/ACM International Conference on Computer-Aided Design*, pages 156–161, 1993.

[SC95]   S. S. Sapatnekar and W. Chuang. Power vs. delay in gate sizing: Conflicting objectives? In *Proceedings of the IEEE/ACM International Conference on Computer-Aided Design*, pages 463–466, 1995.

[SD96]   S. S. Sapatnekar and R. B. Deokar. Utilizing the retiming skew equivalence in a practical algorithm for retiming large circuits. *IEEE Transactions on Computer-Aided Design of Integrated Circuits and Systems*, 15(10):1237–1248, October 1996.

[SEO+02]   S. Sirichotiyakul, T. Edwards, C. Oh, R. Panda, and D. Blaauw. Duet: An accurate leakage estimation and optimization tool for dual-V_t circuits. *IEEE Transactions on VLSI Systems*, 10(2):79–90, April 2002.

[SF94]   T. Soyata and E. G. Friedman. Retiming with non-zero clock skew, variable register and interconnect delay. In *Proceedings of the IEEE/ACM International Conference on Computer-Aided Design*, pages 234–241, 1994.

[SFM93]   T. Soyata, E. G. Friedman, and J. H. Mulligan, Jr. Integration of clock skew and register delays into a retiming algorithm. In *Proceedings of the IEEE International Symposium on Circuits and Systems*, pages 1483–1486, 1993.

[SFM97]   T. Soyata, E. G. Friedman, and J. H. Mulligan, Jr. Incorporating internconnect, register and clock distribution delays into the retiming process. *IEEE Transactions on Computer-Aided Design of Integrated Circuits and Systems*, 16(1):165–120, January 1997.

[She97]   N. Shenoy. Retiming: Theory and practice. *Integration, the VLSI Journal*, 22(1):1–21, January 1997.

[She99]   B. Sheehan. TICER: Realizable reduction of extracted RC circuits. In *Proceedings of the IEEE/ACM International Conference on Computer-Aided Design*, pages 200–203, 1999.

[She02]   B. Sheehan. Library compatible C_{eff} for gate-level timing. In *Proceedings of Design, Automation and Test in Europe Conference*, pages 826–830, 2002.

[SK93]   S. S. Sapatnekar and S. M. Kang. *Design Automation for Timing-Driven Layout Synthesis*. Kluwer Academic Publishers, Boston, MA, 1993.

[SK01]   K. L. Shepard and D.-J. Kim. Body-voltage estimation in digital PD-SOI circuits and its application to static timing analysis. *IEEE Transactions on Computer-Aided Design of Integrated Circuits and Systems*, 20(7):888–901, July 2001.

[SKEW96]   L. M. Silveira, M. Kamon, I. Elfadel, and J. White. A coordinate-transformed Arnoldi algorithm for generating guaranteed stable reduced-order models of RLC circuits. In *Proceedings of the*

IEEE/ACM International Conference on Computer-Aided Design, pages 288–294, 1996.

[SMB96]   V. Singhal, Sharad Malik, and R. K. Brayton. The case for retiming with explicit reset circuitry. In *Proceedings of the IEEE/ACM International Conference on Computer-Aided Design*, pages 618–625, 1996.

[SMCK03]   P. Saxena, N. Menezes, P. Cocchini, and D. A. Kirkpatrick. The scaling challenge: Can correct-by-construction design help? In *Proceedings of the ACM International Symposium on Physical Design*, pages 51–58, April 2003.

[SMO90]   K. A. Sakallah, T. N. Mudge, and O. A. Olukotun. checkTc and minTc: Timing verification and optimal clocking of synchronous digital circuits. In *Proceedings of the IEEE/ACM International Conference on Computer-Aided Design*, pages 552–555, 1990.

[SMO92]   K. A. Sakallah, T. N. Mudge, and O. A. Olukotun. Analysis and design of latch-controlled synchronous digital circuits. *IEEE Transactions on Computer-Aided Design of Integrated Circuits and Systems*, 11(3):322–333, March 1992.

[SN90]   T. Sakurai and A. R. Newton. Alpha-power law MOSFET model and its applications to CMOS inverter delay and other formulas. *IEEE Journal of Solid-State Circuits*, 25(4):584–594, April 1990.

[SNR99]   K. L. Shepard, V. Narayanan, and R. Rose. Harmony: Static noise analysis of deep submicron digital integrated circuits. *IEEE Transactions on Computer-Aided Design of Integrated Circuits and Systems*, 18(8):1132–1150, August 1999.

[SP99]   V. Sundararajan and K. K. Parhi. Low power synthesis of dual threshold voltage CMOS VLSI circuits. In *Proceedings of the International Symposium of Low Power Electronic Devices*, pages 139–144, 1999.

[SPRB95]   V. Singhal, C. Pixley, R. L. Rudell, and R. K. Brayton. The validity of retiming sequential circuits. In *Proceedings of the ACM/IEEE Design Automation Conference*, pages 316–321, 1995.

[SR94]   N. Shenoy and R. Rudell. Efficient implementation of retiming. In *Proceedings of the IEEE/ACM International Conference on Computer-Aided Design*, pages 226–233, 1994.

[SRVK93]   S. S. Sapatnekar, V. B. Rao, P. M. Vaidya, and S. M. Kang. An exact solution to the transistor sizing problem for CMOS circuits using convex optimization. *IEEE Transactions on Computer-Aided Design of Integrated Circuits and Systems*, 12(11):1621–1634, November 1993.

[SS86]   Y. Saad and M. H. Schultz. GMRES: A generalized minimal resid-
            ual algorithm for solving nonsymmetric linear systems. *SIAM
            Journal on Scientific and Statistical Computing*, 7(3):856–869,
            July 1986.

[SS92]   T. G. Szymanski and N. Shenoy.   Verifying clock schedules.
            In *Proceedings of the IEEE/ACM International Conference on
            Computer-Aided Design*, pages 124–131, 1992.

[SS01]   H. Su and S. S. Sapatnekar. Hybrid structured clock network con-
            struction. In *Proceedings of the IEEE/ACM International Con-
            ference on Computer-Aided Design*, pages 333–336, 2001.

[SS02]   S. S. Sapatnekar and H. Su. Analysis and optimization of power
            grids. *IEEE Design and Test*, 20(3):7–15, May-June 2002.

[SSBS92]   N. Shenoy, K. J. Singh, R. K. Brayton, and A. Sangiovanni-
            Vincentelli. On the temporal equivalence of sequential circuits.
            In *Proceedings of the ACM/IEEE Design Automation Conference*,
            pages 405–409, 1992.

[SSE95]   I. Y. Spillinger, L. Stok, and G. Even. Improving initialization
            through reversed retiming. In *Proceedings of the European Design
            and Test Conference*, pages 150–154, 1995.

[SSF95]   H. Sathyamurthy, S. S. Sapatnekar, and J. P. Fishburn. Speed-
            ing up pipelined circuits through a combination of gate sizing and
            clock skew optimization. In *Proceedings of the IEEE/ACM Inter-
            national Conference on Computer-Aided Design*, pages 467–470,
            1995.

[SSP99]   V. Sundararajan, S. S. Sapatnekar, and K. K. Parhi. MARSH:
            Min-area retiming with setup and hold constraints. In *Proceedings
            of the IEEE/ACM International Conference on Computer-Aided
            Design*, pages 2–6, 1999.

[SSP02]   V. Sundararajan, S. S. Sapatnekar, and K. K. Parhi. Fast and
            exact transistor sizing based on iterative relaxation. *IEEE Trans-
            actions on Computer-Aided Design of Integrated Circuits and Sys-
            tems*, 21(5):568–581, May 2002.

[SSWN97]   S. Simon, C. V. Schimpfle, M. Wroblewski, and J. A. Nossek.
            Retiming of latches for power reduction of DSP design. In *Pro-
            ceedings of the IEEE International Symposium on Circuits and
            Systems*, pages 2168–2171, 1997.

[ST83]   T. Sakurai and K. Tamaru. Simple formulas for two- and three-
            dimensional capacitance. *IEEE Transactions on Electron Devices*,
            ED-30(2):183–185, February 1983.

[SVK94]   S. S. Sapatnekar, P. M. Vaidya, and S. M. Kang. Convexity-based algorithms for design centering. *IEEE Transactions on Computer-Aided Design of Integrated Circuits and Systems*, 13(12):1536–1549, December 1994.

[Szy92]   T. G. Szymanski. Computing optimal clock schedules. In *Proceedings of the ACM/IEEE Design Automation Conference*, pages 399–404, 1992.

[TB93]   H. J. Touati and R. K. Brayton. Computing the initial states of retimed circuits. *IEEE Transactions on Computer-Aided Design of Integrated Circuits and Systems*, 12(1):157–162, January 1993.

[TB03]   B. Thudi and D. Blaauw. Non-iterative switching window computation for delay-noise. In *Proceedings of the ACM/IEEE Design Automation Conference*, pages 390–395, 2003.

[TDL03]   S. Tam, U. Desai, and R. Limaye. Clock generation and distribution for the third generation Itanium® processor. In *Digest of Technical Papers, IEEE Symposium on VLSI Circuits*, pages 9–12, 2003.

[TRD+00]   S. Tam, S. Rusu, U. N. Desai, R. Kim, J. Zhang, and I. Young. Clock generation and distribution for the first IA-64 microprocessor. *IEEE Journal of Solid-State Circuits*, 35(11):1545–1552, November 2000.

[TS97]   G. E. Tellez and M. Sarrafzadeh. Minimal buffer insertion in clock trees with skew and slew rate constraints. *IEEE Transactions on Computer-Aided Design of Integrated Circuits and Systems*, 16(4):332–342, April 1997.

[Tsa91]   R.-S. Tsay. Exact zero skew. In *Proceedings of the IEEE International Conference on Computer-Aided Design*, pages 336–339, 1991.

[Tsa93]   R.-S. Tsay. An exact zero-skew clock routing algorithm. *IEEE Transactions on Computer-Aided Design of Integrated Circuits and Systems*, 12(2):242–249, February 1993.

[TSS92]   H. Touati, N. V. Shenoy, and A. L. Sangiovanni-Vincentelli. Retiming for table-lookup field programmable gate arrays. In *FPGA*, pages 89–94, 1992.

[TST+98]   T.-C. Tien, H.-P. Su, Y.-W. Tsay, Y.-C. Chou, and Y.-L. Lin. Integrating logic retiming and register placement. In *Proceedings of the IEEE/ACM International Conference on Computer-Aided Design*, pages 136–139, 1998.

[TTF00]   S. Tsukiyama, M. Tanaka, and M. Fukui. A new statistical static timing analyzer considering correlation between delays. In *Proceedings of ACM/IEEE International Workshop on Timing Issues in the Specification and Synthesis of Digital Systems (TAU)*, pages 27–33, December 2000.

[Vai89]   P. M. Vaidya. A new algorithm for minimizing convex functions over convex sets. *Proc. IEEE Foundations of Computer Science*, pages 332–337, October 1989.

[van90]   L. P. P. van Ginneken. Buffer placement in distributed RC-tree networks for minimal Elmore delay. In *Proceedings of the IEEE International Symposium on Circuits and Systems*, pages 865–867, 1990.

[VBBD96]   P. Vuillod, L. Benini, A. Bogliolo, and G. De Micheli. Clock skew optimization for peak current reduction. In *Proceedings of the International Symposium of Low Power Electronic Devices*, pages 265–270, 1996.

[VC99]   C. Visweswariah and A. R. Conn. Formulation of static circuit optimization with reduced size, degeneracy and redundancy by timing graph manipulation. In *Proceedings of the IEEE/ACM International Conference on Computer-Aided Design*, pages 244–251, 1999.

[VCL+96]   K. Venkat, L. Chen, I. Lin, P. Mistry, and P. Madhani. Timing verification of dynamic circuits. *IEEE Journal of Solid-State Circuits*, 31(3):452–455, March 1996.

[VCMS+99]   A. Vittal, L. H. Chen, M. Marek-Sadowska, K.-P. Wang, and S. Yang. Crosstalk in VLSI interconnections. *IEEE Transactions on Computer-Aided Design of Integrated Circuits and Systems*, 18(12):1817–1824, December 1999.

[VHBM96]   A. Vittal, H. Ha, F. Brewer, and M. Marek-Sadowska. Clock skew optimization for ground bounce control. In *Proceedings of the IEEE/ACM International Conference on Computer-Aided Design*, pages 395–399, 1996.

[VMS95]   A. Vittal and M. Marek-Sadowska. Power optimal buffered clock tree design. In *Proceedings of the ACM/IEEE Design Automation Conference*, pages 497–502, 1995.

[VMS96]   D. Van Campenhout, T. Mudge, and K. A. Sakallah. Timing verification of sequential domino circuits. In *Proceedings of the IEEE/ACM International Conference on Computer-Aided Design*, pages 127–132, 1996.

[VPMS97]   S. V. Venkatesh, R. Palermo, M. Mortazavi, and K. A. Sakallah. Timing abstraction for intellectual property blocks. In *Proceedings of the IEEE Custom Integrated Circuits Conference*, pages 99–102, 1997.

[VR91]   C. Visweswariah and R. A. Rohrer. Piecewise approximate circuit simulation. *IEEE Transactions on Computer-Aided Design of Integrated Circuits and Systems*, 10(7):861–870, July 1991.

[VS94]   J. Vlach and K. Singhal. *Computer methods for circuit analysis and design*. Van Nostrand Reinhold, New York, NY, 2nd edition, 1994.

[vVA+94]   A. van der Werf, J. L. Van Meerbergen, E. H. L. Aarts, W. F. J. Verhaegh, and P. E. R. Lippens. Efficient timing constraint derivation for optimally retiming high speed processing units. In *International Symposium on High Level Synthesis*, pages 48–53, 1994.

[VW93]   C. Visweswariah and J. A. Wehbeh. Incremental event-driven simulation of digital FET circuits. In *Proceedings of the ACM/IEEE Design Automation Conference*, pages 737–741, 1993.

[Wag88]   K. D. Wagner. Clock system design. *IEEE Design and Test of Computers*, pages 9–27, October 1988.

[WBL+94]   N. Wehn, J. Biesenack, T. Langmaier, M. Munch, M. Pilsl, S. Rumler, and P. Duzy. Scheduling of behavioral VHDL by retiming techniques. In *Proceedings of the European Design Automation Conference*, pages 546–551, 1994.

[WC02]   T.-Y. Wang and C. C.-P. Chen. 3-D thermal-ADI: A linear-time chip level transient thermal simulator. *IEEE Transactions on Computer-Aided Design of Integrated Circuits and Systems*, 21(12):1434–1445, December 2002.

[WCJ+98]   L. Wei, Z. Chen, M. Johnson, K. Roy, and V. De. Design and optimization of low voltage high performance dual threshold voltage CMOS circuits. In *Proceedings of the ACM/IEEE Design Automation Conference*, pages 489–494, 1998.

[WCR+99]   L. Wei, Z. Chen, K. Roy, Y. Ye, and V. De. Mixed-Vth (MVT) CMOS circuit design methodology for low power applications. In *Proceedings of the ACM/IEEE Design Automation Conference*, pages 430–435, 1999.

[WE93]   N. Weste and K. Eshraghian. *Principles of CMOS VLSI Design*. Addison-Wesley, Reading, MA, 2nd edition, 1993.

[WR93]   U. Weinmann and W. Rosenstiel. Techology mapping for sequential circuits based on retiming techniques. In *Proceedings of the European Design Automation Conference*, pages 318–323, 1993.

[WV98]   Q. Wang and S. B. K. Vrudhula. Static power optimization of deep submicron CMOS circuits for dual V_t technology. In *Proceedings of the IEEE/ACM International Conference on Computer-Aided Design*, pages 490–496, 1998.

[Wya85]   J. L. Wyatt. Signal delay in RC mesh networks. *IEEE Transactions on Circuits and Systems*, 32(5):507–510, May 1985.

[Wya87]   J. L. Wyatt, Jr. Signal propagation delay in RC models for interconnect. In A. E. Ruehli, editor, *Circuit Analysis, Simulation and Design*, volume 2. North-Holland, Amsterdam, The Netherlands, 1987.

[XBG+01]   T. Xanthopoulos, D. W. Bailey, A. K. Gangware, M. K. Gowan, A. K. Jain, and B. K. Prewitt. The design and analysis of the clock distribution network for a 1.2 GHz Alpha microprocessor. In *Proceedings of the IEEE International Solid-State Circuits Conference*, pages 402–403, 2001.

[XD97]   J. G. Xi and W. W.-M. Dai. Useful-skew clock routing with gate sizing for low power design. *Journal of VLSI Signal Processing Systems for Signal, Image and Video Technology*, 16(2/3):163–180, June/July 1997.

[XK95]   T. Xue and E. S. Kuh. Post routing performance optimization via multi-link insertion and non-uniform wiresizing. In *Proceedings of the IEEE/ACM International Conference on Computer-Aided Design*, pages 575–579, 1995.

[YKK95a]   H. Yotsuyanagi, S. Kajihara, and K. Kinoshita. Resynthesis for sequential circuits designed with a specified initial state. In *Proceedings of the IEEE VLSI Test Symposium*, pages 152–157, 1995.

[YKK95b]   H. Yotsuyanagi, S. Kajihara, and K. Kinoshita. Synthesis for testability by sequential redundancy removal using retiming. In *International Symposium on Fault Tolerent Computing*, pages 33–40, 1995.

[YMP+01]   H. Yalcin, M. Mortazavi, R. Palermo, C. Bamji, K. A. Sakallah, and J. P. Hayes. Fast and accurate timing characterization using functional information. *IEEE Transactions on Computer-Aided Design of Integrated Circuits and Systems*, 20(2):315–331, February 2001.

[YS90]   H. Youssef and E. Shragowitz. Timing constraints for correct per-
           formance. In *Proceedings of the IEEE/ACM International Con-
           ference on Computer-Aided Design*, pages 24–27, 1990.

[Zho03]   H. Zhou. Timing analysis with crosstalk is a fixpoint on a complete
           lattice. *IEEE Transactions on Computer-Aided Design of Inte-
           grated Circuits and Systems*, 22(9):1261–1269, September 2003.

[ZS98]   M. Zhao and S. S. Sapatnekar. Timing optimization of mixed
           static and domino logic. In *Proceedings of the IEEE International
           Symposium on Circuits and Systems*, 1998.

Index